UNDERSTANDING BIOINFORMATICS

UNDERSTANDING BIOINFORMATICS

By

Dr. Rajeev Tyagi

Deptt. of Chemistry
M.M.H. Post Graduate College
Ghaziabad (U.P.)
(India)

DISCOVERY PUBLISHING HOUSE PVT. LTD.
NEW DELHI-110 002

Published by:
Tilak Wasan
DISCOVERY PUBLISHING HOUSE PVT. LTD.
4831/24, Ansari Road, Prahlad Street
Darya Ganj, New Delhi-110002 (India)
Phone: +91-11-23279245, 43764432
Fax: +91-11-23253475
E-mail: parul.wasan@gmail.com
info@discoverypublishinggroup.com
discoverypublishinghouse@gmail.com
web: www.discoverypublishinggroup.com

***First Edition:* 2011**
ISBN: 978-81-8356-855-5

Understanding Bioinformatics

Printed in:
INDIA

PREFACE

The present title *"Understanding Bioinformatics"* provides a structured approach to learning by covering all the important topics in a uniform, systematic format. The book has been comprehensively designed incorporating recent advances in this fast moving field. It is written to provide accessible information on bioinformatics in compact form for undergraduate students in biology and related life sciences. It will be useful for both beginning students and those who are more advanced. In addition, busy lecturers who require a quick reference compendium will find it useful, particularly for tutional planning. Simple, yet hopefully clear figures and tables are provided throughout the book.

The over-riding goal of this book, and indeed of the whole *Understanding series,* is to present the essential information concerning microbiology in a compact, readily accessible form which leads itself to student learning and revision. The convergence of various approaches has generated a rich panorama of detail, the significance of which we are still attempting to unraval. The present text has been written as an introduction to this rapidly growing field.

To make the work more comprehensive and informative, the author has consulted many authoritative books, research journals, abstracts, monographs etc., so there can be no claim to originality except in the manner of treatment.

The author expresses his thanks to his friends and colleagues whose continue inspirations have initiated him to bring out this book.

The author expresses his gratitude to Mr. Wasan and staff of M/s Discovery Publishing House Pvt. Ltd. for their whole hearted co-operation in the publication of this book.

In the mean time, the author will remain sincerely responsible for any shortcomings of the book and be grateful to the readers for their suggestions and constructive criticism for the continuous betterment of the book. He takes this opportunity to appeal to the readers to send their suggestions straightaway to his Publisher.

Author

Contents

1

INTRODUCTION

Every research worker would like to have the tools on hand to make his job quicker and more efficient, and with the advent of the World Wide Web many of the tasks associated with molecular biology have become freely available online. In the past when a scientist wanted to know something about a particular subject then the first option was to talk to colleagues in the laboratory and ask for their advice. If that was not sufficient then it was off to the library to scan abstracts or the latest journals for the relevant information.

However times are changing and so are working habits. Why ask questions from people in your laboratory when you can ask the same question on the Bionet newsgroups http://www.bio.net from research workers all over the world? Why thumb through textbooks for references when you can type in keywords to an Internet search engine such as Lycos or Alta Vista and get a satisfactory answer in no time at all? But often you find that the major search engines index everything on the Web, which makes it difficult to find exactly what you want. So often it is more profitable to use search engines that are totally dedicated to biology.

In Europe you could use BiowURLd or Bio-Hunt which deal exclusively with biology-related subjects. Another comprehensive listing exists at the Virtual Library in the BioSciences division. http://www.vlib.org/Biosciences.html, and from China there is the NEE-HOW project, http://biology.neehow.org which is an invaluable resource for research workers from the Pacific rim.

In the USA one of the original and best lists of Biological resources, put together by Keith Robison can be found at Harvard http://golgi.harvard.edu/biopages.list and of course there is the ever popular *Pedro's BioMolecular Research Tools*. If you are looking specifically for software related to bioinformatics, then there is the BioCatalogue or if you are looking for an obscure database then there is DBcat from Infobiogen the EMBnet node in France.

There are also a few good newsletters, which deal specifically with what is happening in the world of bioinformatics. EMBnet produces a quarterly newsletter, which gives an update of the latest developments at the different EMBnet nodes throughout Europe.

The EBI has its own industry programme and they produce a newsletter called the *Bioinformer*. One special feature within this newsletter is the BioEvents Calendar, that allows people to advertise workshops, conferences, or symposiums. In the USA there are two major newsletters associated with bioinformatics. The NCBI news-letter is at http://www.ncbi.nlm.nih.gov/VVeb/Newsltr/index.html and the National Centre for Genomic Research, the NCGR newsletter, is at http://www.ncgr.org/ ncgr/ncgr_newsletter.html

Changing Face of Networking

In the early 1990s academic research workers had the networks all to themselves. Today however, the demography of those using the networks and their reasons for using the networks have completely changed. The competition for bandwidth is fierce between the commercial and academic sector.

The Georgia Institute of Technology has been conducting user surveys on the use of the Internet since 1994. Over a four-year period there have been many radical changes in attitude towards the use and abuse of the Web. The most recent surveys show that when it comes to using the WWW the two main activities that people engage in are collecting personal information, and using the Web purely for entertainment. The academic no doubt will be distressed that work and education only occupy equal third place. Academics are no longer the only people using the internet and they may feel that their research work suffers because of the 'info-tourists' on the web. Gone are the days when the only people on the network were scientists with Unix boxes. More and more people are coming online from home and the humble PC seems to have cornered the market.

In the past scientists relied on centralized systems, with systems administrators installing and maintaining programmes. Nowadays since the installation of many programmes on PC's has been fully automated, scientists are doing it for themselves. This means that the scientist

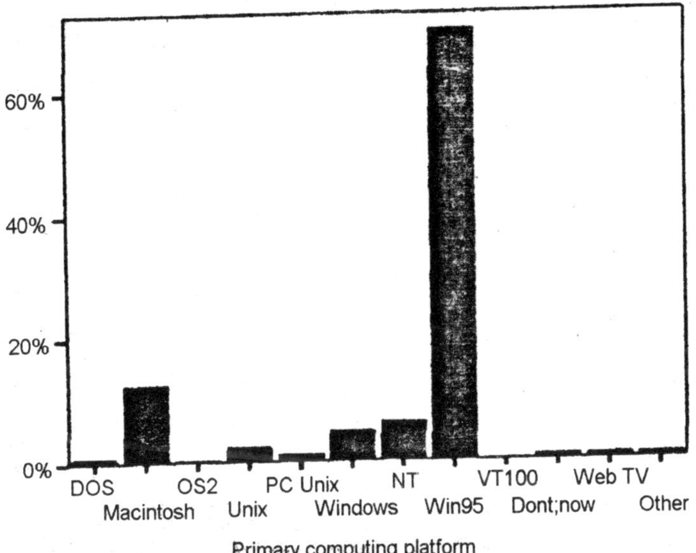

Fig. 1.1. Primary computing platform.

needs to be aware of the trends that are driving the internet forward. Applications will be written for platforms that are being used the most. If the scientist insists that they can get by with their VT100 terminal and a text based Lynx browser very soon they will be unable to browse sites that are visually rich or rely on Java scripts or corba interfaces. It is clear from the latest survey results that the most used widely used computing platform is Windows 95. No doubt this is partly due to the popularity of Microsoft Internet Explorer which comes bundled with the operating system. The browser wars between Netscape and Microsoft have already led to legal battles in the American courts.

Networking in Europe

When it comes to networking not all countries are created equal. The EMBnet organization has developed a service called 'Network Performance monitoring in EMBnet'. This project has monitored the efficiency of networking throughout Europe between the EMBnet nodes.

In 1995, a *Network Usage and Quality Advisory* Group of the Dutch Network organisation SURFnet, defined 'an upper RTT limit of 125 msec. *without packet loss'* as a minimum QOS (*Quality of Service*) level for interactive online work. The KIT values from the Dutch Embnet node to the EBI can be represented by the following graph.

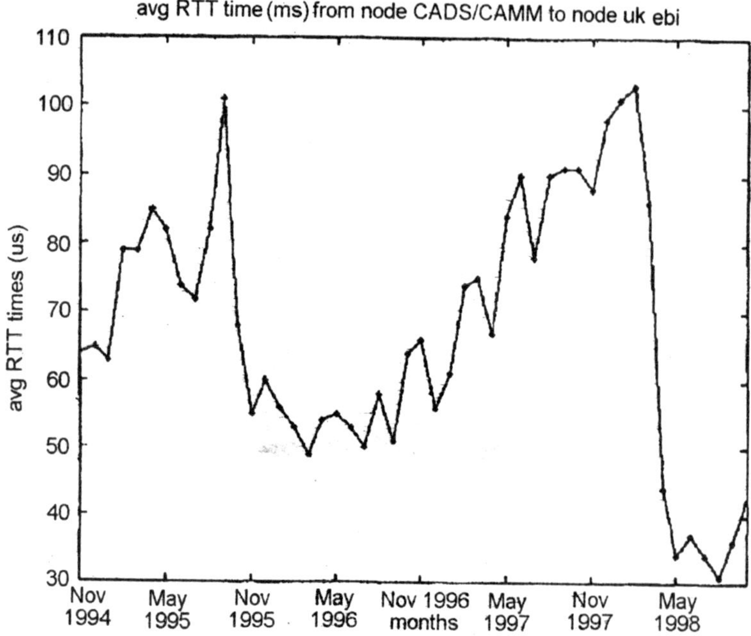

Fig. 1.2. Average round trip times in ms from EmbNet node CAOS/CAMM in the Netherlands to the EBI in the UK.

This shows that the RTT from the Netherlands to the EBI in the UK has consistently been below the recommended time of 125 msec, which means scientists from the Netherlands, should have no difficulties in contacting the EBI web server. It is interesting to note that the results from October 1998 show that of the thirty-two nodes monitored, twenty-two have a WIT of less than 125 msec. This must surely be good news for networking within Europe.

It is essential that research workers learn to use the services provided for them within their own countries. Penalties are always paid when you network across international borders. It would seem that the more borders you cross, the less efficient the network becomes. However, networking within your own country is more efficient because more often than not a basic infrastructure already exists between the major universities. In the mind of the molecular biologist, however, Mecca is either at the NCBI or EBI and that is the direction they religiously point their browsers to, only to suffer frustration when they cannot get their work done due to poor bandwidth and increased traffic directed towards these sites. For this reason EMBnet tries to co-ordinate their activities so that all the EMBnet nodes provide easy access for database query and retrieval. Many of the EMBnet nodes use a mirror package to update their databases on a daily basis, via remote ftp from the databases stored on the EBI anonymous ftp server at ftp://ftp.ebi.ac.uk/pub/databases.

The major databases such as EMBL or SWISS-Prot are then indexed at the EMBnet nodes and can be queried with the SRS package. SRS which was developed at EMBL Heidelberg by Thure Etzold, has been adopted by many of the EMBnet nodes throughout Europe and also abroad. SRS is also unique in that it is able to index very many different databases.

E-mail Servers for Sequence Retrieval

Networking for the biologist has a very short history and many of the services developed in the eighties are still in use today. Indeed for people with very bad network connections the use of E-mail servers is still the preferred method of obtaining sequences or running homology similarity searches such as FASTA or Blast. The main depositories for sequence data are are found in the UK at the European BioInformatics Institute (EBI), and at the National Centre for Biotechnology Information (NCBI) in the United States. In addition these two institutes collaborate with the DDJBB in Japan.

Both the EBI and NBCI run e-mail servers that will allow you to retrieve sequences via e-mail. To obtain information on how to run the e-mail server at EBI you simply send a e-mail message and include in the main body of the message the word help and full instructions will be sent via e-mail on how to operate the service.

A similar method for sequence retrieval is employed by the NCBI and the e-mail query system utilizes the Entrez retrieval system that they have developed for their website. Many people would argue that getting sequence via e-mail is old-fashioned technology. It is primitive in that it only delivers simple ascii-formatted text. However the e-mail query server at the NCBI is clever enough to be able to return the sequence to you in a variety of different formats including GenBank, FASTA, or Html. It is often more convenient to shoot off a query by e-mail and get an answer within a few minutes than it is to struggle with trying to access a website that has bandwidth problems. To receive full instructions on how the server works just send an e-mail message and in the main body of the message type the word help. I have often found that people who have used an e-mail server generally have a better understanding of databases and sequence retrieval than those who have only used a WWW interface.

Table 1.1. A list of the EMBNET nodes and some other sites around the world which support SRS

WEHI, Melbourne, Australia
Vienna Biocenter EMBnet Node, Vienna, Austria
Belgian EMBnet Node (BEN), Brussels, Belgium
DBBM-IOC, Fiocruz, Rio de Janeiro, Brazil
CBR-NRC, Halifax, Canada The Genome Mine, Base4 Bioinformatics, Canada
CBI EMBnet Node, University of Beijing, China
CSC, Otaniemi, Espoo, Finland
INFOBIOGEN, Villejuif, France Institut Pasteur, Paris, France
LBMRPM INRA/CNRS, Auzeville, Toulouse, France
DKFZ, Heidelberg, Germany
EMBL, Heidelberg, Germany
GBF, Braunschweig, Germany
MIPS—MPG/GSF, Martinsried/Munich, Germany
Bioinformatics Centre, University of Pune, India
INCBI EMBnet Node, Dublin, Ireland
Weizmann Institute BCD, Rehovot, Israel
CNR EMBnet Node, Bah, Italy
CRISCEB, Second University of Naples, Italy
IVR, Kyoto University, Japan
Biotek EMBnet Node, Oslo, Norway
IBB-PAS EMBnet Node, Warsaw, Poland
IGC EMBnet Node, Oeiras, Portugal
SRCG, Novosibirsk, Siberia, Russia
BIC, National University Hospital, Singapore
CNB EMBnet Node, Madrid, Spain
Biomedical Centre (BMC), Uppsala, Sweden
ExPASy, Geneva, Switzerland
CAOS/CAMM Center, Nijmegen, The Netherlands
RIGEB-MRC, Gebze, Kocaeli, Turkey
Adlib, CAB International, Wallingford, UK
EMBL-EBI, Hinxton, Cambridge, UK
HGMP-RC, Hinxton, Cambridge, UK
MBDC Oxford, Oxford University, UK
SEQNET EMBnet Node, Daresbury, UK
Sanger Centre, Hinxton, Cambridge, UK
IUBio, Indiana University, USA

Similarity Searches via E-mail

The two most popular e-mail servers dealing with similarity searches are Blast from the NCBI. and FASTA from EBI. For help regarding these e-mail servers you can send an e-mail message and complete instructions on how to formulate an e-mail message to be processed by these servers will be returned to you via e-mail. Again it should be stressed that once you understand how to compose an e-mail message to submit a Blast query via E-mail, then you can be more discriminating when you are asked to repeat the procedure via the WWW. As it is most people just opt for the default parameters and never experiment with different options.

The BLAST algorithm was developed by the National Center for Biotechnology Information at the National Library of Medicine. The BLAST family of programs employs this algorithm to compare an amino acid query sequence against a protein sequence database or a nucleotide query sequence against a nucleotide sequence database, as well as other combinations of protein and nucleic acid. It used to be that the NCBI exclusively provided access to BLAST but in recent years you can now run BLAST searches from many different sites around the world, which is a clear indication that this programme has become a very popular method for doing homology searches. The fact that it appears in so many places may be due to the fact that it is available for free from the NCBI anonymous ftp server. Historically the EBI has always provided homology searches through FASTA.

Speed Solutions for Similarity Searches

In recent years there has been an increase in the use of specialized hardware for doing similarity searches. Four companies in particular have pioneered this approach, and the turn around time for running a search against the whole of Swiss-Prot has been reduced to around 10 seconds using the Smith-Waterman algorithm.

Time logic

Time Logic from the USA has introduced DeCypher Bioinformatics Accelerators and they have implemented a number of algorithms namely, Gapped BLAST2 PSI-BLAST, Affine Smith-Waterman, FrameScaich, ProfileSearch, ProfileScan, and ClustalW with graphical rendition of dendrogram (Java applet).

Compugen

Compugen have succeeded in introducing the Biocellorator to many pharmaceutical companies to aid them in their search for new and novel drugs. The EBI has a biocellorator, which is online and is available for public use. At the EBI there are two different interfaces to this service. The one provided by Compugen called GeneWeb and a simple custom interface developed at the EBI. The interface to the BIC-SW at the EBI is very compact and easy to use. Most people just accept the default settings, paste in their query sequences and run the program. If you are from the Mediterranean area then perhaps it would be more convenient to try the GeneWeb interface from the Weizmann Institute in Israel, which is also open to the public for unregistered users.

Paracel FDF

At the Swiss EMBnet node you can find the Paracel Fast Data finder (FDF), which is designed to help bioinformatics departments dramatically increase the rate at which they can find high-scoring potential genornic targets. Paracel claim that GeneMatcher is the first commercially available genetic data analysis system to use custom ASIC technology that can analyse similarities or differences in DNA or protein sequences up to 1000 times faster than traditional computer systems. Competition to discover novel genes is of great interest to pharmaceutical companies because if it is possible to identify just one critical target gene then this can result in an application for a patent on a product. Therefore any method that combines speed with sensitivity is a very valuable tool in the hands of the research worker.

Sequence Retrieval via the WWW

If you are in a country with a poor Internet connection then working with E-mail servers for the retrieval of sequences is often the best option. However. there are many excellent servers in different parts of the world and they should not be ignored. even if you do live on the other side of the planet. Your geographical location should be the first consideration when accessing a remote site. It is best to access a site that is in close proximity. Two of the most popular services for sequence retrieval are Entrez from the NC:BI and SRS from the EBI. However there are other options available and if you are in the Pacific rim area then it might be worthwhile to look at the services offered by DDBj in Japan or Maestro service from National Centre for Genomic Research (NCGR) on the West Coast of the USA.

Entrez from the NCBI

The NCBI is the only place in the world where you will find the Entrez service and it concentrates on a few databases namely, nucleotide sequences, amino acid sequences, 3-D structures, Genomes, Taxonomy, and Literature-PubMed. One of its strengths is that it provides access to PubMed and this is a key factor in its popularity and success. Effective August 3, 1998, NLM implemented a system enhancement that dramatically increases the speed of the system. This redesign of the way PubMed stored and retrieved information will improve users search time—a search that previously took approximately 18 seconds to run in PubMed now runs under 2 seconds.

In Entrez you select the database you wish to query, for example the protein database and then you are allowed to string a number of keywords together, like 'Rhizobium Ausubel nodulation' and those entries that meet the criteria will be displayed.

SRS from the EBI

SRS is a very powerful tool for querying databases and it would seem to be the preferred querying system within Europe. You select a database and fill in your search criteria as keywords.

SRS will then display two hits in Swiss-Prot for that particular year with a sequence range between 500 and 700. It should also be noted that SRS also gives the possibility to launch an application such as BLAST or PASTA for any of the sequences that you care to select. You may also select different views of a sequence. For example the PASTA format, which then allows you to launch a multiple sequence alignment using ClustalW, directly from within SRS. This method is a great time saver since there is no need to cut and paste your sequence into a separate ClustalW application.

SUBMITTING SEQUENCES

Not only does the research worker want to query, retrieve and analyse sequences, occasionally they also want to submit their own sequences to the databanks be it GenBank or EMBL. The three major organizations that collect sequence information work in collaboration with each other so that sequences entered into GenBank are transferred daily by FTP to both EBI and DDBJ (and vice versa) in an attempt to keep the major databases synchronized.

At any given time the three institutes are continually swapping data so it is a false idea to believe that any one database is more current than the other. All three institutes have online methods of submitting sequence data through the Web. The NCBI were the first to come online with BANKIT. The EBI then followed with WEBIN and the Japanese at DDBJ have Sakuara.

It should also be noted that the NCBI developed a stand-alone programme for MAC's, PC's, and Unix called SEQUIN that allows the end-user to enter their data from a personal computer and to send the submission via e-mail or to simply post the disk to the appropriate institute where it is then uploaded into the database. Sequin is strongly recommended if you have bulk submissions to make.

Bankit at NCBI

Bankit is convenient for quick submission of sequence data to the NCBI. Banklt allows you to enter sequence information into a form, edit as necessary, and add biological annotation (e.g. coding regions, mRNA features). Banklt transforms your data into GenBank format for you to review and when your record is completed, it can be submitted directly to GenBank. You have the option of adding information by using text boxes to describe in your own words the source of the sequence and its biological features. The GenBank annotation staff reviews the submitted textual information, incorporates it into the appropriate structured fields, and returns the record by e-mail for your review.

Sequin from NCBI

Sequin is stand-alone program for the MAC, PC/Windows and UNIX. Sequin is an interactive, graphically oriented program based on screen forms and controlled vocabularies that guide you through the process of entering your sequence and providing biological and bibliographic annotation. Sequin is designed to simplify the sequence submission process and to provide graphical viewing and editing options. This program is optimal for submitting multiple sequences, mutation studies, phylogenetic sets, population sets, and segmented sets. It incorporates robust error checking and accommodates very long sequences and complex annotations.

Although Sequin has been implemented by the NCBI, the opening screen allows you to select which database you would like to submit you sequence to be it GenBank, EMBL, or DDBJ. Usually when a sequence is submitted there may be a process whereby the submitter has to be in contact with the annotators of the sequence by telephone to clarify certain details. Therefore it is wise to choose a submission centre in your geographical region if you want to avoid long distance telephone calls.

Once you have completed the submission depending on which database you have selected at the beginning you will be prompted to send an e-mail to NCBI, EMBL. Sequin runs on Macintosh, PC/ Windows, and UNIX computers. The program itself, along with its on-line help documentation, is available by anonymous FTP from the EBI (UN) or from the *NCBI* (USA).

Webin from EBI

The EBI WWW tool (Webin) guides the user through a sequence of WWW forms allowing the user to submit sequence data and descriptive information in an interactive and easy way. All the information required to create a database entry will be collected during this process:

(a) Submitter information.

(b) Release date information.

(c) Sequence data, description, and source information.

(d) Reference citation information.

(e) Feature information (e.g. coding regions, regulatory signals etc.).

Data submissions are usually processed within two working days of receipt and the authors are sent notification of their accession number(s). Authors will be asked whether their submitted data can be made available to the public immediately or whether they should be withheld until an author-specified date. Data are never withheld after publication.

Once a database entry has been created from a submission, a copy is sent to the submitter for their reference and for comments or corrections. However, it often happens that the entry is correct when it is created but, with the passage of time, becomes out of date. The authors may make corrections to the sequence itself, or may discover new features of the sequence. Since such findings are often not published, the only way to keep entries correct and up to date is if the authors communicate their new findings to the database. At the EBI this can be done by completing an update form available from the Anonymous FTP.

A new service that has been instituted at EBI is scanning for vectors before submitting your sequence. You are able to check your sequence data prior to submission for potential vector contamination by running a BLASTN search against EMVEC, a vector database containing information on more than 2000 vectors from the EMBL/ GenBank/DDBJ Database SYN(thetic) division. The results will list sequences producing significant alignments and associated information like vector name, score, alignment, etc. The EBI suggests that you remove vector contamination from your sequence data before submitting to the database.

Sakura from DDJB

SAKURA is a web-based DNA data submission system for DDBJ. You can select either the English or Japanese version, However, data input must be done in English only, regardless of language version selected. SAKURA allows you to save your document before completion and submit multiple sequences sequentially.

Historically there has been a collaboration between EBI, NCBI, and DDBJ. These three sites are still the only places that have the infra-structure set up to handle the submission of nucleotide sequences to the databases, be they EMBL or Genbank or DDBJ. For this reason they are also looked upon as the only places where you can do queries and retrieval, or perform homology searches, or multiple sequence alignments. This is no longer true and with the advent of EMBnet, many of the national nodes are able to supply services that are not offered by the major centres. These three major centres have a policy of making all of their databases publicly available, and when distributed network of databases exists in many different parts of the globe then it can only be for the benefit of molecular biologists worldwide.

2

DNA MICROARRAYS

Gene expression, which includes two processes, namely transcription of data from a DNA template to RNA, and translation, involving the construction of proteins on the basis of the information about the linear sequence of their amino acids encoded in the RNA, lies behind all functions of cells in organisms. The idea of DNA microarray technology is to monitor gene expression processes by measuring levels of RNA species in biological samples, for example tissue cells or blood. RNA molecules in the samples are labeled by using appropriate techniques and presented to an array of spots, where complementary-DNA (cDNA) fragments corresponding to known coding DNA sequences are placed. It is also possible to copy RNA sequences, by a reverse transcription mechanism, back to a DNA strand, label the DNA with fluorescent dyes and hybridize it to a complementary-DNA probe fixed on the microarray. The measurement of RNA levels is based on the fundamental property of nucleotide sequences, already mentioned in this book, of binding (hybridizing) to their complements. If the level of the RNA product corresponding to the DNA placed at a spot X in the microarray is high in the sample being analyzed then we should observe a high fluorescence signal at spot X. The complementary-DNA sequences placed in the spots of microarrays are designed and synthesized on the basis of our the knowledge about the content of genomes of organisms. The number of spots in a DNA microarray is comparable to the number of known genes in the organism studied, and can reach tens of thousands in DNA microarrays dedicated to the human genome. Appropriate technology allows the precise positioning and stabilizing of complementary-DNA probes on glass or plastic plates and then, after hybridization of the labeled target molecules, estimation of the level of RNA in the sample by reading the intensity of the signals from the dots of the DNA array.

Experiments with DNA microarrays are performed to help study issues in biology and clinical practice, regarding cellular mechanisms, the functions of genes and proteins, the structure of gene networks and pathways, relating the risk of being affected by diseases to gene expression profiles, etc. Gene expression profiling has been successfully used in many medical research programs concerning monitoring cellular process, measuring the response of cells or tissues to therapeutic agents, classification or detection of disease symptoms and many other problems.

Gene expression experiments lead to the creation of huge data sets consisting of tens of thousands of RNA species, corresponding to known or putative genes. However, most of the genes whose expression profiles are generated in microarray experiments may be unrelated to the processes or phenomena being studied. Therefore an appropriate methodology must be developed for inference based on gene expression levels obtained from DNA microarrays to filter out irrelevant information. The amount of data makes it intractable manually, and appropriate bioinformatic tools must be applied to review and organize the information. The mathematical modeling approaches used must be consistent (i) with the aim of the study, i.e., they must help in verifying the hypotheses behind the experiment, and (ii) with the specific character of microarray data. A study based on gene expression usually involves analyzing issues such as the following:

1. Are there differences between the gene expression profiles obtained in experiments *A* and *B*?
2. Is gene *X* overexpressed or underexpressed in experiment *A* versus experiment *B*?
3. Is there a correlation between the gene profiles in experiments *A* and *B*?
4. Is there a group of genes that is always overexpressed (or underexpressed) under the experimental conditions of *A*?
5. If there is an environmental, temporal, or spatial factor behind the experiment, are there genes that follow the pattern of this factor?

An important aspect of the analysis of microarray data is the possibility of repeating the experiment or of collecting multiple samples under different experimental conditions or situations. Some issues, such as (iv) or (v) above can be efficiently resolved when experiments are repeated many times, but become very difficult otherwise.

The process of inferring useful information from expression profiles involves several steps where models and mathematics, along with heuristics and intuition, are necessary for choosing between algorithms and between values of numerous parameters. In this chapter, we review some approaches to the analysis of large data sets consisting of gene expression profiles and illustrate them using publicly available data sets obtained from DNA microarrays. We present some basic techniques for data normalization, and the related statistics of expression levels and logarithms of expression levels. We present a maximum likelihood method for modeling probability distributions of logarithms of expressions, and we show how it can be applied to infer useful information.

Modeling probability density functions by Gaussian mixtures allows one to study clustering properties that arise from similar values or patterns of change of gene expressions. Mixture model analysis can be enhanced by incorporating information about repetition of the measurement into the construction of the likelihood function. We overview some methods of dimensionality reduction. We also demonstrate or mention methods for class prediction and class discovery, namely hierarchical clustering, the K-means method, and linear and nonlinear classifiers. By analyzing distances and grouping data, hierarchical clustering explores the wealth of information encoded in the positive or negative correlations of expression values induced by simultaneous increases or decreases in gene expression.

Design of DNA Microarrays

The two main techniques for the measurement of gene expression measurement are (i) complementary-DNA arrays and (ii) oligonucleotide arrays. In both cDNA and oligonucleotide microarrays, labeled (dyed) target RNA or DNA molecules bind to immobilized complementary-DNA probes. The cDNA technology is cheaper and it is possible to implement and develop it on a laboratory scale. Therefore there are many cDNA standards, and cDNA chips dedicated to many research programs are manufactured by genomic laboratories and small enterprises, as well as university laboratories. The oligonucleotide technology is more involved and expensive and there are only a few industrial manufacturers of microarray chips and scanners that are used in scientific research; the most widely known is *Affymetrix*.

In the cDNA technology, DNA strands (100–5000 base long) are presynthesized and placed on a glass or plastic plate by microrobots called *DNA arrayers* or spotters. The probes are deposited on the plate by a method similar to inkjet printing. The surface of the plate itself must be prepared appropriately to allow attraction and stabilization of the probes. The sequence of DNA probes is established using data from the DNA databases. RNA material from experimental samples is isolated and then reverse transcribed to DNA with the use of reverse transcriptase. The cDNA strands obtained are labeled with fluorescent dyes and then presented to the DNA array, where a hybridization process occurs. Usually, two different fluorescent dyes (Cy3, orange or green, and Cy5, dark red) are used to label the samples. One color is used for the case samples, corresponding to a biological experiment, and another color is used for the reference or control RNA samples. The case and control samples, labeled with different dyes, are mixed and presented to the cDNA on the microarray spots, where the hybridization process takes place. The effective information obtained in a cDNA experiment is the fluorescence intensities, measured with the use of a scanning device.

Owing to imperfect making of cDNA probes, reading the dye intensities is prone to errors from many sources, such as nonuniform intensities of the dyes in the spots and irregularities in the spot shapes. In order to reduce the influence of these errors, digital image-processing techniques are applied as a preprocessing stage in data acquisition. Images of cDNA arrays are often published to accompany the results of their analysis. Therefore they are widely available on the Internet. The study in the paper referred to was devoted to identifying types of B-cell lymphomas (human lymphatic cancers) on the basis of the cDNA gene expression profiles. The cDNA microarrays were specially designed and included 17856 cDNA clones chosen from DNA libraries related to B cells and lymphomas. In order to achieve the aim of the study, both malignant and normal tissue and blood samples were collected and analyzed with the use of the manufactured cDNA microarrays. Normally, cDNA microarray apparatuses are equipped with computer software that allows for the intensities of the fluorescence signals in the cDNA spots to be translated into numerical values. Using that program; one can estimate the fluorescence intensities of both the red Cy5 and the green Cy3 spectral components.

Oligonucleotide microarrays are composed of DNA strands synthesized in situ on the solid support. The oligonucleotide probes have a length of 20–50 bases and each gene is represented by 20–25 probes corresponding to its exonic (coding) fragments. The technology of assembling oligonucleotide microarrays is a combination of solid-phase chemical synthesis and photolithography technology similar to that used for manufacturing LSI and VLSI electronic circuits. In order to control the growth of the oligonucleotide strand, chemical reagents are applied to block reactive groups on the bases and on the deoxyribose ring (benzoyl and isobutyryl are used to block the reactive NH_2 groups on the bases and dimethoxytrityl is used to block the 5′ position on the deoxyribose ring).

The blocking reagent at 5′ position of the deoxyribose ring is removed before the phase of adding the new base to the DNA sequence. In the process of manufacturing an oligonucleotide array, tens of thousands of different oligonucleotide strands are synthesized at the same time on a plate. This is achieved by steps where access to spots on the plate is opened and closed by the use of light-activated photolithographic masks, synchronized with the stages of synthesis of solid-phase single-strand oligonucleotide DNA. By using light-sensitive masks, oligonucleotide spots are either masked or made available to the reagents, with a time schedule controlled by a computer algorithm.

The targets for the immobilized probes on the microarray plate are fluorescent, labeled RNA molecules. After hybridization, the signal intensity is measured with the use of a digitally controlled laser-optic system. The signal corresponding to one gene results from averaging and comparing signals from many probes; the value available to the user is proportional to the level of the RNA species. Some controls, or flags, are also added that report on the quality of the signal.

The basic difference between the two microarray formats is in the lengths of the complementary-DNA probes. In oligonucleotide arrays the probes are of constant length in the range of 20–50 base pairs, while in cDNA arrays the lengths of the cDNA strands differ between spots. In oligonucleotide microarrays, in principle, the readouts of fluorescence intensities are comparable between different spots. In contrast, in cDNA microarrays, the different probe lengths change the reaction rates between spots. Therefore experiments performed using cDNA always require the samples to be compared to controls. Each spot containing DNA clones is exposed both to targets from the sample RNA and from the control RNA. As already mentioned, the case samples and control samples are labeled with fluorescent dyes of different colors (red and green), and their RNA levels are compared by measuring the intensities of the corresponding dyes. In the Affymetrix oligonucleotide microarrays, one complementary-DNA probe is also associated with two spots called PM and MM, where PM stands for "*perfect match*" and MM for "*mismatch*". The PM probe is a DNA strand that corresponds uniquely to a gene. The MM probe has one nucleotide in the middle altered. It is intended that PM versus MM comparison will increase the precision of measurements by eliminating errors resulting from cross-bindings and background hybridization.

Kinetics of the Binding Process

The hybridization reaction can be represented by the following scheme:

$$R + L \underset{k_r}{\overset{k_f}{\rightleftarrows}} C,$$

where R denotes the number of oligonucleotide strands available for reaction, L is the molar concentration of free target RNA samples, and C stands for the number of bound complementary complexes. The coefficients k_f and k_r are the forward (binding) and reverse (unbinding) reaction rates. The units for the rates are mol^{-1} time^{-1} for k_f and time^{-1} for k_r. Different units of measurement are used for different molecules: L is measured as a molar concentration, while R and C are numbers of molecules. The different units underline the nature of the experimental setup. If we denote the molar volume of the solution interacting with the probe by V and the Avogadro's number by N_A, then we can compute the number of free target RNA molecules in the solution as LVN_A. The kinetics of the process of hybridization is governed by the law of mass action. The rate of forward binding of target molecules to immobilized probes is proportional to the product of the concentrations of the target strands and free, probes and the unbinding process is a first-order reaction with a rate proportional to C. This results in the following balance of flows:

$$\frac{dC}{dt} = k_f RL - k_r C. \quad \ldots(1)$$

We assume that at the beginning, i.e., at $t = 0$, there are no hybridization complexes, i.e., $C(0) = 0$, and we use the notation $L(0) = L_0$ and $R(0) = R_T$, where the subscript T stands for the total number of oligonucleotides available for hybridization. Since one RNA strand binds to one oligonucleotide, resulting in one binding complex, the following equalities for the flows hold

$$\frac{dR}{dt} = VN_A \frac{dL}{dt} = -\frac{dC}{dt},$$

which results in $R(t) + C(t) = R_T$ and $VN_AL(t) + C(t) = VN_AL0$. So $L(t)$ and $R(t)$ can be expressed in terms of $C(t)$ and substituted in leading to

$$\frac{dC}{dt} = k_f[R_T - C(t)]\left[L_0 - \frac{C(t)}{N_AV}\right] - k_rC(t)$$

or

$$\frac{dC}{k_f(R_T - C)(L_0 - C/N_AV) - k_rC} = dt.$$

The left-hand side of the above equation can be expanded into two first-order fractions, which allows one to find the analytical solution. Often it is possible to approximate the above dynamics to only one exponent. Specifically, one of the following two asymptotic situations may hold. (i) There is a large excess of particles in the immobile probe over the potential number of binding targets, i.e., $R_T \gg C$, or $(R_T - C)/R_T \approx 1$, which results in

$$C(t) = \frac{R_TL_0}{K_D + R_T/N_AV}\left[1 - \exp\left(-\frac{t}{\tau R}\right)\right], \qquad \text{...(2)}$$

where $K_D = k_r/k_f$ and

$$\tau_R = \frac{1}{k_f(K_D + R_T/N_AV)}.$$

(ii) There is a large excess of free RNA strands with respect to the number of binding complexes, i.e., $L_0 \gg C/N_AV$, or $(L_0 - C/N_AV)/L_0 \approx 1$, which leads to

$$C(t) = \frac{R_TL_0}{K_D + L_0}\left[1 - \exp\left(-\frac{t}{\tau_L}\right)\right], \qquad \text{...(3)}$$

$$\tau_R = \frac{1}{k_f(K_D + L_0)}.$$

Microarray experiments are very often planned in such a way that there is a large excess of particles in the immobile probe over the potential number of binding targets, i.e., case (i) holds. From (eq. 2) one can see that, after equilibrium has been reached, or at a predefined instant of time, the intensity of the fluorescence signal measured at a microarray spot is proportional to the level of the corresponding RNA species in the analyzed sample, i.e., $C \sim L_0$.

Along with the process of binding labeled target RNA or DNA molecules to their corresponding complementary-DNA sequences, a processes of cross-hybridization may occur. Cross-hybridization is the binding of target molecules to non-corresponding DNA regions. It differs from complementary hybridization in its coefficients k_f and k_r. Dai et al. have noted that the forward hybridization coefficients k_f have comparable values for complementary hybridization and cross-hybridization, whereas the coefficient k_r of the reverse process is at least of one order of magnitude, higher for cross-hybridization than for complementary hybridization. Therefore in a first-approximation model, the influence of cross-hybridization on the measured values of gene expression can be neglected.

Data Preprocessing and Normalization

Owing to imperfections in the assembly processes and the high density and large number of spots on DNA microarray chips, measurements of expression levels are contaminated, to a substantial extent, by measurement noise. Also, there is a large variation in the concentration levels of different RNA macromolecules resulting from their biological functions in cells. Signals are not present or are low at some of the microarray spots owing to the absence of the corresponding RNA species in the analyzed sample. On the other hand, some of the RNA molecules that appear in very small amounts are nevertheless crucial for the proper functioning of many molecular mechanisms. Therefore, it is important to employ preprocessing and normalizing steps on the raw expression data, with the aim of eliminating some errors and reducing the variation of the measurement noise. The preprocessing and normalization procedures are based on the hypothesis that there is a systematic error between experiments, which can be removed (or reduced) by averaging or scaling. The aims of the normalization steps are (i) to label measurements of low reliability, which introduce mostly noise accompanied by a very low or no useful signal, and (ii) to estimate of the useful signal by averaging or applying other transformations of this type to the components of the measurements. Several possible approaches to reducing the variation of errors are mentioned in the literature. They can be grouped into two classes: (i) normalization based on a model of the transcription process, such as normalization to the total or ribosomal RNA, normalization to housekeeping genes, normalization to a reference RNA, or normalization by spiked in control RNA sequences; and (ii) normalization by applying empirical scaling functions that transform the distributions of the expression values to the desired shape. We can also distinguish between normalization applied to one scan of a microarray chip and normalization resulting from averaging over repetitions of an experiment under the same or similar conditions.

Despite the preprocessing steps built into the microarray software by their manufacturers, the researcher often needs to add additional rules for data analysis. When analyzing data from several DNA microarray chips, one may encounter the situation where a gene is marked present on one chip and absent on another one. Also, for some microarray standards, it may happen that the values of gene expression returned by the microarray software are negative. Some rules must then be introduced to infer useful information from data of these types. After this, data on the expression of genes in microarrays are often visualized by plotting histograms of RNA levels over different subsets of the genes spotted on the plate.

DNA microarray experiments involve both measuring and comparing RNA levels between samples taken from cell lines at different times, between different cell lines, between different individuals, and so forth. So, the statistical description of the interplay between random and systematic elements in the samples may be complex and may require a considerable research effort.

Even in very carefully planned biological experiments, many sources of error are rather poorly recognized. Therefore, despite their potential advantages, models of transcription processes are often ignored, and, instead, methods that belong to class (ii) above are used to reduce the systematic errors. The operations performed on microarray data include logarithmic transformations (or more general nonlinear transformations), centering transformations, and variance standardization transformations. These transformations allow one to eliminate some systematic errors without bothering with precise models of the mechanisms that cause them. The approaches that belong to class (ii) can be called *black-box* modeling, since no (or almost no) hypotheses are introduced regarding the hybridization process or its parameters.

An important issue in the analysis of DNA microarray data is standardization of the data processing procedures, which paves the way towards comparing microarray experiments between different studies. One element of ensuring repeatable results is to standardize the normalization procedures. If the studies to be compared use the same normalization procedures, their results should be comparable at the level of the final estimates of RNA levels. However, despite the existence of some recommendations regarding normalization methods, there is still no single standard for these methods. Therefore, a possibility, often applied when experimental data are published, is making available fluorescence intensities signals at the level of the probes. This makes it possible to redo statistical analyses from the raw data to the final conclusions and allows easier comparisons between different studies. The normalization procedures published in the literature are supported by a lot of publicly available software that performs normalization procedures for microarrays.

Generally, it is commonly believed that the preprocessing and normalization steps are very important and have an impact on the overall results of studies involving the use of DNA microarrays. The lack of an adequate normalization step can lead to misleading conclusions. Below we describe some approaches to normalization procedures for DNA microarrays.

Normalization Procedures for Single Microarrays

First we describe the normalization procedure, included as a part of the software developed by manufacturer of DNA microarrays scanner, Affymetrix. This procedure can be applied to a single microarray. The algorithm for pre-processing and normalizing the measurements of RNA intensities is being modified as products are upgraded; here we present the version called MAS 5.0. The files of expression level data produced by DNA microarray scanners are organized such that the measured expression levels are marked additionally by labels that describe their level of reliability. In the DNA microarray chips manufactured by Affymetrix, each gene is represented by K probes (also called *spots*) placed on the surface of the plate. For example, in Affymetrix human-genome chip HG U133, $K = 11$, L the length of the cDNA sequence = 25, and the number of genes is 22000. Each probe contains a pair of two cDNA sequences denoted by PM (Perfect Match) and MM (Mismatch). The PM sequence is a sequence L bases long sampled from the exonic part of the gene. MM sequence is equal to the PM sequence at all bases except the 13th (the one in the center) which is altered to another base. The PM and MM signals from the K probes are used to decide whether the measurement should be labelled P for present, A for absent or M for marginal. To make the decision, discrimination scores R_k, $k = 1,2, \ldots K$ are computed with the use of the following formula:

$$R_k = \frac{PM_k - MM_k}{PM_k + MM_k}$$

where PM_k and MM_k, $k = 1,2,\ldots, K$ are the intensities of the PM and MM fluorescence signals, respectively. By definition, the discrimination scores R_k are in the interval (–1, 1). The null hypothesis is that the gene is absent, which can be denoted by $H_0 = H_{Absent}$. This means that there are no RNA strands in the sample that correspond to the probe cDNA sequence, and we should expect fully random binding of RNA sequences to both PM and MM. The alternative hypothesis is that the gene is present, i.e., $H_A = H_{Present}$. In this situation the PM spot should attract more RNA strands than the MM spot, owing to its higher affinity. To decide between H_{Absent} and $H_{Present}$, or, more precisely, to accept or reject $H_0 = H_{Absent}$, a rank statistic is used. The values $R_1, \ldots, R_{11}$ are assigned a plus or minus sign by the criterion that a plus is assigned if $R_k >$ Threshold, otherwise a minus is assigned. Here Threshold is a predefined number; the default is Threshold = 0.015. Then, the positive values of R (those to which a plus sign has been assigned) are sorted and ranked in ascending order, the negative values of R are sorted in descending order, and the one-side Wilcoxon signed- rank statistical test is applied, to compute the detection p-values. Although the rank test has a slightly lower power than the paired t- test, its advantage is robustness against non normality. Commonly, when statistical tests are applied, the critical value for p is taken as $p_{critical} = 0.05$. A similar, but slightly modified approach is adopted here. The following three subintervals are used to categorize the expression of genes on the basis of the computed p-values:

$$\text{if } p \in \begin{cases} (0, 0.04) \text{ then the gene is labeled } P \text{ (Present)}, \\ (0.04, 0.06) \text{ then the gene is labeled } M \text{ (Marginal)}, \\ (0.06, 1) \text{ then the gene is labeled } A \text{ (Absent)}. \end{cases} \quad \ldots(4)$$

The final estimate of the expression level of the gene is computed from component measurements PM_k and MM_k, $k = 1,2,\ldots, 11$, as follows:

$$X = \frac{1}{\#S} \sum_{k \in S} (PM_k - MM_k), \quad \ldots(5)$$

where S is the subset of probes for which $PM_k - MM_k$ is within three standard deviations of the average.

To illustrate the labeling P, M, and A, let us look at the analysis of the data, available electronically at an accompanying Web site. The experimental study in that paper is based was based on Affymetrix microarrays of type Hu6800, with 7129 spots, including cDNA strands corresponding to 6817 human genes and a number of control spots corresponding to control RNA species that could be used for calibration purposes. Gene expression profiles were obtained from patients affected by two types of leukemia, acute lymphoblastic leukemia (ALL) and acute myeloid leukemia (AML). The initial training set included 38 patients (27 ALL and 11 AML). The measurements of RNA levels were labelled P, M, and A according to the decision rule (eq. 4) and assigned numerical values based on differences between PM and MM intensities as described in equation (eq. 5). In total, the training set contained 38 × 7129 = 270902 measurements (spots). Among them there were 187892 spots labelled absent A, 78632 spots labelled present P and 4378 spots labelled marginal M. So, it seems that majority of spots contain no useful data. However, we must take into account the fact that the data are not 270902 independent spots but, rather, 7129 spots each repeated 38 times. Typically, if there are multiple measurements (experiments) of the expression for each gene, then the genes labeled A in all experiments are excluded from further analysis.

Normalization Based on Spiked-in Control RNA

Normalization methods which belong to class (i) above, "normalization based on a model of the transcription process", need information or modification of the design of the experiment and/or knowledge about the properties of the cDNA sequences deposited on the array spots. This involves using spiked-in control RNA species. Spiked-in control RNA species are RNA strands of known, controlled levels in the probes, with base sequences that correspond to sets of spots used in the DNA microarray chips. Let us assume that M spiked-in control RNA species have been added to each of N oligonucleotide microarray chips in a biological experiment and that each sample contains the same level of spiked-in control RNA number i. The matrix of the $M \times N$ measurements of spiked-in control RNA expression levels is

$$\begin{bmatrix} x_{11} & x_{12} & \ldots x_{1N} \\ x_{21} & x_{22} & \ldots x_{2N} \\ \vdots & \vdots & \ddots \vdots \\ x_{M1} & x_{M2} & \ldots x_{MN} \end{bmatrix},$$

where xij stands for the expression level of the ith spiked-in control in the jth microarray chip. The basic hypothesis, following from the design of the experiment, is that changes in the expression levels of the ith spiked control between chips are a consequence of the systematic errors resulting from the chip-manufacturing process. We assume the following multiplicative model for these data:

$$x_{ij} = m_i \cdot r_j \cdot e_{ij}. \qquad ...(6)$$

By m_i, we denote the true expression level of the *i*th spiked-in control. The *j*th microarray chip is characterized by a multiplicative modifying factor r_j and e_{ij} is a random multiplicative error. After applying the logarithmic transformation, y_{ij} = log(x_{ij}), μ_i = log(m_i), ρ_j = log(r_j), ε_{ij} = log(e_{ij}), the model (eq. 6) becomes additive:

$$y_{ij} = \mu_i + \rho_j + \varepsilon_{ij}. \qquad ...(7)$$

We now assume that ε_{ij} is normally distributed with zero mean and a variance σ_2^i . In other words, we make one more assumption, that the variance of the logarithm of the error ε_{ij} does not depend on the array number, *j*, but only on the spiked-in control species number, *i*. In the model (eq. 7), the y_{ij} are observations and μ_i, ρ_j, σ_i are parameters to be estimated. Under the hypothesis of independence between different spiked-in controls and between errors in different chips, the following log-likelihood function is associated with the data:

$$L = \log \prod_{i=1}^{M} \prod_{j=1}^{N} p(y_{ij}, \mu_i + \rho_j, \sigma_i^2) \qquad ...(8)$$

where *p*(.) is the normal probability density function

$$p(y_{ij}, \mu_i + \rho_j, \sigma_i^2) = \frac{1}{\sqrt{2\pi\sigma_i^2}} \exp\left[\frac{(y_{ij} - \mu_i - \rho_j)^2}{\sigma_i^2}\right]. \qquad ...(9)$$

Substituting (eq. 9) in (eq. 8) and taking derivatives with respect to the parameters μi, ρ_j and σ_i leads to the system of equations

$$\hat{\mu}_i = \frac{1}{N} \sum_{j=1}^{N} (y_{ij} - \hat{\rho}_j), \qquad ...(10)$$

$$\hat{\rho}_j = \frac{\sum_{i=1}^{M} (\hat{\sigma}_i^2)^{-1} (y_{ij} - \hat{\mu}_i)}{\sum_{i=1}^{M} (\hat{\sigma}_1^2)^{-1}}, \qquad ...(11)$$

$$\hat{\sigma}_i^2 = \frac{1}{N} \sum_{j=1}^{N} (y_{ij} - \hat{\mu}_i - \hat{\rho}_j)^2. \qquad ...(12)$$

After the above system of equations has been solved, $\hat{\mu}_i, \hat{\rho}_j$ and $\hat{\sigma}_i$ are the maximum-likelihood estimates of the means and variances. The form of (eq. 10)(eq. 12) dictates an easy method for their solution by iteration. One problem, which arises in the case of combinations or mixtures of continuous distributions is that for some data, the maximization of (eq. 8) may diverge to infinity and, in iterations we will have $\hat{\mu}_i - \hat{\rho}_j \to y_{ij}$ and $\hat{\sigma}_i^2 \to 0$. A method to avoid this is to introduce a prior distribution of $\hat{\sigma}_i^2$ such that the probability density is zero at $\hat{\sigma}_i^2 = 0$, i.e., $p(\hat{\sigma}_i^2 = 0) = 0$. Hartemink et al. assumed that the prior distribution for joint distribution of variances was a Wishart distribution, which, with additional assumptions about symmetry and independence, leads to a joint probability density of $\hat{\sigma}_1^2, \hat{\sigma}_2^2, \ldots, \hat{\sigma}_M^2$ in the factorized form

$$p_W(\hat{\sigma}_1^2, \hat{\sigma}_2^2 \ldots, \hat{\sigma}_M^2) = \prod_{i=1}^{M} C(\alpha, t)\left(\frac{1}{\hat{\sigma}_i^2}\right)^{(\alpha-3)/2} \exp\left(-\frac{t}{2\hat{\sigma}_i^2}\right), \quad \ldots(13)$$

where α and t are predefined constants and $C(\alpha, t)$ is an appropriate scaling factor. A plot of the family of functions

$$p(\sigma^2) = \left(\frac{1}{\sigma^2}\right)^{(\alpha-3)/2} \exp\left(-\frac{t}{2\sigma^2}\right), \quad \ldots(14)$$

with $\alpha = 5$ and several choices for t, is presented. From the plot and (eq. 14), we see that the distribution in (eq. 13) satisfies $p_W(\hat{\sigma}_i^2 = 0) = 0$, provided that $t > 0$. Maximization of the log-likelihood

$$L_1 = \log \prod_{i=1}^{M} C(\alpha, t)\left(\frac{1}{\hat{\sigma}_i^2}\right)^{(\alpha-3)/2} \exp\left(-\frac{t}{2\hat{\sigma}_i^2}\right) \prod_{j=1}^{N} p(y_{ij}, \mu_i + \rho_j, \sigma_i^2)$$

instead of the L given by (eq. 8), leads to equations (eq. 10) and (eq. 11); (eq. 12) is replaced by

$$\hat{\sigma}_i^2 = \frac{\sum_{j=1}^{N} (y_{ij} - \hat{\mu}_{ij} - \hat{\rho}_j)^2 + t}{N + \alpha - 3}. \quad \ldots(15)$$

It can be verified that the likelihood L_1 does not diverge to infinity and that the iterations following from (eq.10), (eq.11), and (eq.15) (such that (eq.12) is replaced by (eq.15)) cannot result in $\hat{\mu}_i - \hat{\rho}_j \to y_{ij}$, $\hat{\sigma}_i^2 \to 0$.

From the estimate $\hat{\rho}_j$ computed from (eq.10)(eq.12) there follows the optimal scaling factor for the jth chip, denoted by s_j,

$$s_j = \frac{1}{\hat{r}_j} = \exp(-\hat{\rho}_j) = \prod_{i=1}^{M} \left(\frac{\hat{m}_i}{x_{ij}}\right)^{w_i}$$

where $\hat{m}_i = \exp(\hat{\mu}_i)$ and the weights w_i (weights) computed as follows:

$$w_i = \frac{(\hat{\sigma}_i^2)^{-1}}{\sum_{k=1}^{M}(\hat{\sigma}_k^2)^{-1}}.$$

Finally, we conclude that all expression levels in chip number j should be multiplied by the factor s_j.

Methods similar to that presented above can also be used for normalization based on other data from a microarray, for example, the expression of house-keeping genes, which should remain at constant levels across experiments.

RMA Normalization Procedure

An often applied method for the normalization of results of DNA microarray experiments is robust multiarray analysis (RMA. This approach proceeds in several steps and assumes a model for the expression profiles analogous to that presented in (eq.6) and (eq.7). It can be used both for single measurements and for normalization based on repetition of experiments on measurements of gene expression in DNA microarrays. In order to apply RMA normalization, one must have access to the data from separate probes in the microarray measurement.

We assume that a microarray scanning experiment has been repeated J times, and introduce the notation

$$PM_{ijk} \text{ and } MM_{ijk},$$

where, as in (eq.5), PM and MM stand for reads of the perfect-match and mismatch intensities, and the indices are used as follows: the index $i = 1,2,\ldots, I$ represents the genes in the samples, the index $j = 1, 2,\ldots, J$ represents different samples, and the index $k = 1,2, \ldots, K$ represents the number of the probe. A characteristic feature of RMA is that it relies only on the perfect match probe intensities PM_{ijk}. The RMA algorithm proceeds in the following steps.

Background correction

The following exponential–normal model is assumed for PM_{ijk}

$$PM_{ijk} = bg_{ijk} + s_{ijk}, \qquad \ldots(16)$$

where bg_{ijk} stands for the background noise and s_{ijk} represents the hybridization signal. The background noise bg_{ijk} results from optical noise and non-specific binding and is modeled by a normal distribution with an array-specific mean level $E(bg_{ijk}) = \beta_i$. The hybridization signal s_{ijk} is assumed to be distributed exponentially. Using these assumptions a filter is constructed which is intended to reduce the background noise. This filter uses the model of a mixture of normal and exponential distribution.

Quantile normalization

Quantile normalization between different microarrays is applied to the background-corrected hybridization signals. This equalizes quantile-quantile plots and histograms of expression profiles. Quantile normalization of two samples x and y of equal length involves:

1. Sorting both x and y in decreasing order, which leads to vectors x_{sorted} and y_{sorted}. Sorting is an appropriate renumeration, so we can write

$$x_{sorted} = \text{renum}_x(x),$$

and

$$y_{sorted} = \text{renum}_y(y).$$

2. Computing the mean of x_{sorted} and y_{sorted}

$$z = \frac{1}{2}(x_{sorted} + y_{sorted});$$

and

3. Computing the quantile-normalized vectors by an operation of restoring the original order, applied to the mean z. By restoring the original order we mean applying the inverse operators renum_x^{-1} and renum_y^{-1} to the vector z, i.e.,

$$x_{quantile\text{-}normalized} = \text{renum}_x^{-1}(z),$$

and

$$y_{quantile\text{-}normalized} = \text{renum}_y^{-1}(z).$$

Quantile normalization of more than two samples is defined in the way analogous to the above. A variant of quantile normalization based on replacing the operator mean by the operator median is also possible.

Additive model for normalization

For the signals Y_{ijk} defined as the background-corrected, quantile-normalized and log-transformed values of PM_{ijk}, we now use the following additive model:

$$Y_{ijk} = y_{ij} + \alpha_{ik} + \varepsilon_{ijk}$$

where y_{ij} is the final normalized log transformed expression level, α_{ik} is the probe affinity effect, and ε_{ijk} is random Gaussian noise. The above model is fitted to the preprocessed data and the final, normalized values of the logarithms of the expression signals, y_{ij}, are obtained.

Correction of Ratio–Intensity Plots for cDNA

An example of normalization of cDNA microarray data based on a model of the transcription process is a procedure based on ratio–intensity plots. This procedure leads to corrections to the values of the logarithm of the ratio R/G. The method uses the hypothesis that the statistics of the log-ratio $\log(R/G)$ should be independent of the logarithm of the intensity product, $\log(RG)$, where R is the fluorescence intensity of the red dye and G is that of the of green dye. For these data R is denoted by *Ch*2 and G by *Ch*1. A scatterplot of the logarithms of the products of intensities $\log_{10}(R \cdot G)$ versus the logarithms of the ratios $\log_2(R/G)$ is presented on the left. Data points are $[\log_{10}(R_k \cdot G_k), \log_2(R_k/G_k)]$, where k ranges from 1 to the number of clones. Each data point is represented by a plus sign. A systematic bias is clearly seen. We have estimated this bias by means of a third-order polynomial relation, i.e.,

$$Y_k = F(X_k) = a_0X^3{}_k + a_1X^2{}_k + a_2X_k + a_3 \qquad \text{...(17)}$$

where $Y_k = \log_2(R_k/G_k)$, $X_k = \log_{10}(R_k \cdot G_k)$, and the above equation is understood in the least squares sense. The parameters a_0, a_1, a_2 and a_3 can be estimated by using a simple least-squares algorithm. The resulting estimated bias relation is plotted as a black curve. Removing the bias, i.e., subtracting the estimated bias from the logarithms of ratios, which means taking $Y_{k\ \text{corrected}} = Y_k - F(X_k)$, leads to normalized (corrected) values of $\log_2(R_k/G_k)$. A plot of $\log_{10}(R_k \cdot G_k)$ versus $\log_2(R_k/G_k) - F(\log_{10}(R_k \cdot G_k))$ (corrected), is shown in the right-hand part of Figure. The simple version of the least-squares method, which we used to remove the bias in the data does not take into account the non-uniform density of data points along the $\log_{10}(R \cdot G)$ axis. A more adequate statistical analysis would be the use of a locally weighted linear regression to remove the ratiointensity bias.

Statistics of Gene Expression Profiles

The common paradigm in inference based on gene expression profiles is that information on the process behind the DNA microarray experiment is encoded in the ratios of RNA levels between different experiments. The researcher is interested in the factor by which the RNA concentration has increased or decreased from one measurement to another. In order to change the ratios to a linear scale, a logarithmic transformation is applied as a preprocessing stage. This approach has been confirmed by the statistics of gene expression. Histograms of RNA levels, or their ratios resemble an exponential distribution. After the logarithmic transformation transform distributions become similar to normal, or can be modeled by several normal components.

The typical statistics of RNA levels in oligonucleotide DNA microarrays can be illustrated well by using again the data. From the training set consisting of 38 patients (27 ALL and 11 AML), we have chosen one ALL patient (sample number 1) and one AML patient (sample number 38). We have excluded erroneous (negative) measurements of expression. The histogram bars are scaled as relative frequencies; in other words, their areas add up to one. In the upper plots in Fig. 2.1 the values on the horizontal axis are the RNA levels, and in the lower plots, base-2 logarithms of the RNA levels. One can see that the probability density functions in the upper plots resemble exponential functions and their logarithmic transforms are normal-like.

More precisely, distributions of the logarithms of RNA levels are bimodal (or multimodal) and so they should be modeled by mixtures of two or more normal distributions rather than by a single normal distribution. We can also illustrate, analogously to the above, the statistics of the data obtained from cDNA microarray chips. Again

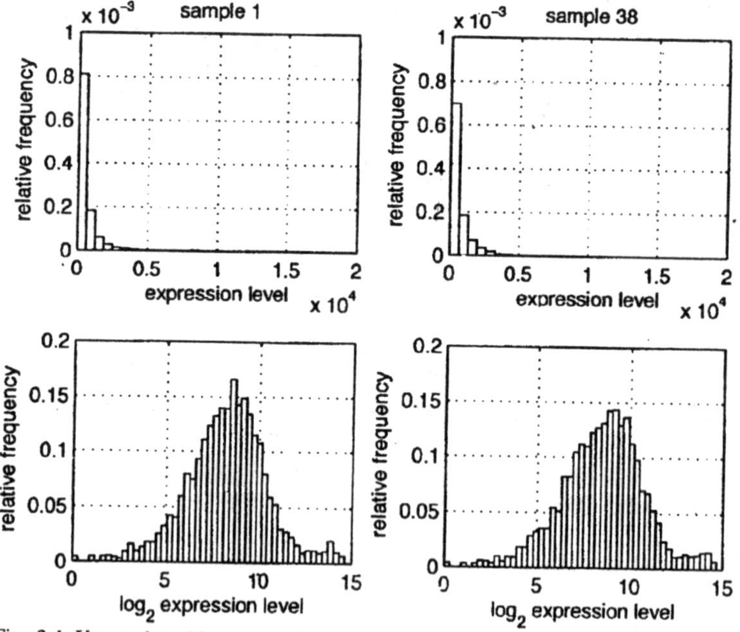

Fig. 2.1. Upper plots: histograms of expression levels for sample 1 and sample 38 from the data. Lower plots: corresponding histograms of logarithms to base 2 of the expression levels.

we use, as an example, the expression profile of the sample *lc7b023*. Signals corresponding to the RNA levels, denoted by *Ch*1 and *Ch*2, are the background-corrected mean fluorescence intensities of pixels within the ellipses of the spots, computed according to the following equations:

$$Ch1 = CH1I - CH1B$$

$$Ch2 = CH2I - CH1B$$

where the index 1 means the green (532 nm) and the index 2 means the red (635 nm) component of the spectrum, *CH1I* and *CH2I* are the fluorescence intensities averaged over the pixels of the spot, and *CH1B* and *CH2B* are the average fluorescence intensities of the background pixels. Replacing the mean values *CH1B* and *CH2B* by medians can lead to better robustness of the measurements against noise. The data file created for the measurements on the cDNA chip contains quality control parameters for each spot specifying, for example, number of pixels in the spot with intensities greater than background. These parameters allow on to test statistical hypotheses related to the presence or absence of gene expression products in the sample. Spots which generated measurements that did not allow rejection of the hypothesis

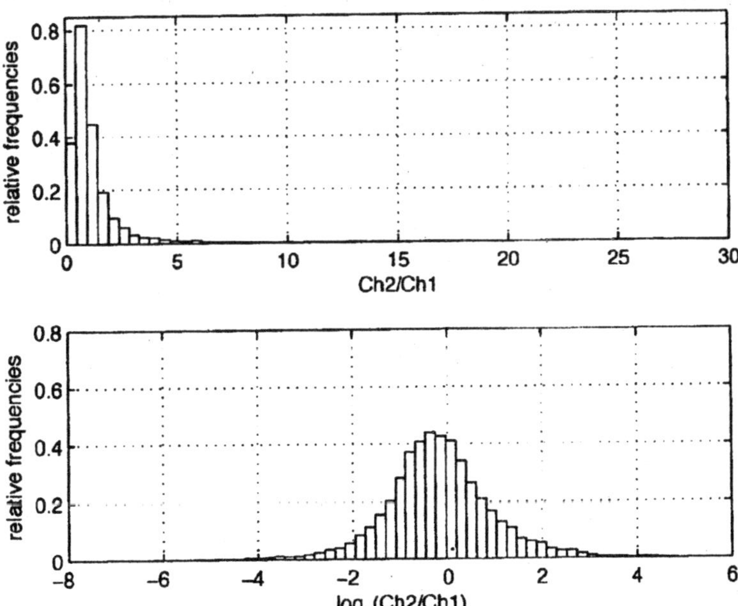

Fig. 2.2. Histograms of the ratios Ch2/Ch1 and their logarithms log_2(Ch2/Ch1) for the dataset lc7b023.

of the absence of the corresponding gene or which led to values of *Ch*1 or *Ch*2 less than zero, are removed from the analysis. The basic signals used in our inference based on these cDNA microarray experiments are the ratios *Ch*2/*Ch*1. In Fig. 2.2 histograms of the ratios *Ch*2/*Ch*1 (in the upper plot) and their logarithms $\log_2$ (*Ch*2/*Ch*1) (in the lower plot) are presented. Again the distribution of the ratio *Ch*1/*Ch*2 is exponential-like, while, after logarithmic transformation the distribution of $\log_2$(*Ch*2/*Ch*1) is similar to normal.

Modeling Probability Distributions of Gene Expressions

We now focus our attention on the probability distribution of the logarithms of gene expression levels. We denote the logarithm of the expression level of the *n*th gene by x_n. From the plots in Figs. 2.1 and 2.2 it can be seen that although they are similar to normal probability density functions, the logarithms of expression levels from have more than one mode. Therefore it seems reasonable to model such distributions by mixture densities.

Several researchers have proposed using mixtures of distributions to solve various issues in the interpretation of DNA microarray data. The paper [189] overviews methods for obtaining parameters of mixtures of different distributions and proposes the use of mixtures of factor analyzers for unsupervised classification of colon and leukemia gene expression datasets. In a decomposition of expression level probability density functions into Gaussian components is used to set thresholds to classify expression levels as "change", "no change", "overexpressed", "under- expressed", etc. Thresholds for logarithms of gene expression levels and for SLRs (logarithms of ratios of expression levels) are often set intuitively. For example, a gene is considered overexpressed if the base-2 logarithm of its expression level exceeds the average by 1, or, equivalently its expression is twice as much as the average. However, when gene expressions are classified into groups, for example, "no change", "overexpressed" and "underexpressed", based on boundaries computed by the use of estimated probability density functions corresponding to different clusters, the results are often biologically more sound. This can be verified by using information on ontologies of genes. In study, different variants of Bayesian-mixture based clustering procedures were studied. In a hierarchical agglomerative clustering method was used for an initial guess of the mixture parameters and the mixture model was combined with a dimensionality reduction technique for the analysis of cutaneous melanoma data. In a mixture model was used to determine the differential expression of genes in the presence of mixed cell populations. In mixture modeling was applied to the problem of missing measurements in DNA microarrays.

The approach of using probability density function decomposition to grouping genes into coexpressing clusters can also be used successfully in the case where DNA microarray experiments are conducted to compare gene expression profiles under different experimental conditions. In that case it seems natural to impose the requirement that genes remain clustered into the same component over all experiments. However, the components can change their parameters between different experimental conditions. Also, it is possible that the repetitive structure represented by (eq.22) may be combined with variability resulting from changing experimental conditions. Decomposition into components according to different patterns of expression can be used, for example, to study the functions of genes involved in the cell cycle. In decomposition into Gaussian mixture was used to study time-course data on gene expression profiles of cancer cells after irradiation.

The probability distribution of a normal mixture model, $p(x)$, is given by the formula below:

$$p(x) = \sum_{k=1}^{K} \alpha_k p_k(x) \qquad \text{...(18)}$$

In the above equation x denotes the natural logarithm of the expression level; α_k, $k = 1,2,\ldots, K$, are weighting coefficients $\Sigma_{k=1}^{K} \alpha_k = 1$; and $p_k(x)$, $k = 1,2,\ldots K$, are probability density functions of normal components, i.e.,

$$p_k(x) = p_k(x, \mu_k, \sigma_k) = \frac{1}{\sqrt{2\pi\sigma_k}} \exp\left[-\frac{(x-\mu_k)^2}{2\sigma_k^2}\right], \qquad \text{...(19)}$$

where μ_k is the expectation and σ^2_k is the variance of the kth normal component. The parameters to be adjusted are the number of components, the expectations and variances of each component, and the weighting coefficients.

They can be estimated by the maximum likelihood method. Denoting by x_n the logarithm of the expression level of the nth gene, we express the likelihood function as follows:

$$L(x_1,\ldots,x_N) = \prod_{n=1}^{N} \sum_{k=1}^{K} \alpha_k p_k(x_n). \qquad \text{...(20)}$$

To estimate the parameters, $\alpha_1, \alpha_2, \ldots, \alpha_K, \mu_1, \mu_2, \ldots, \mu_K, \sigma_1, \sigma_2, \ldots, \sigma_K$, we can use the iterations of the EM algorithm.

A decomposition of the form (eq.18) is natural for modeling the probability density function for one microarray measurement experiment. However, it does not incorporate information obtained by measurement repetition. When gene expression measurements are repeated, it becomes natural to make the assumption that all measurements of the expression of one gene belong to the same component of the mixture distribution. Let the number of repetitions be R, and denote by

$$\overline{x}_n = [x_n^1 \; x_n^2 \; \ldots \; x_n^R]$$

the vector of the logarithms of the expression obtained in repeated measurements, for gene number n. We have used an overbar in order to distinguish more explicitly non repeated and repeated measurements. Under the hypothesis that all repeated measurements of one gene belong to one Gaussian component, the probability density function for $\overline{x}_n$ becomes

$$p(\overline{x}_n) = \sum_{k=1}^{K} \alpha_k \prod_{r=1}^{R} p_k(x_n^r, \mu_k, \sigma_k), \qquad \ldots(21)$$

where $p_k(x, \mu_k, \sigma_k)$ is the pdf of the normal distribution given in (eq.19). The likelihood function analogous to (eq.20) but accounting for repeating measurements has the form

$$L(\overline{x}_1, \ldots, \overline{x}_N) = \prod_{n=1}^{N} \sum_{k=1}^{K} \alpha_k \prod_{r=1}^{R} p_k(x_n^r, \mu_k, \sigma_k) \qquad \ldots(11.22)$$

where R stands for the number of repeated measurements. The likelihood (eq.22) can be maximized by techniques analogous to those discussed before. The EM iterations leading to an increase of the likelihood function (eq.22) assume the following form:

$$p(k|\overline{x}_n, p^{old}) = \frac{\alpha_k^{old} \prod_{r=1}^{R} p_k(x_n^r, \mu_k^{old}, \sigma_k^{old})}{\sum_{k=1}^{K} \alpha_k^{old} \prod_{r=1}^{R} p_k(x_n^r, \mu_k^{old}, \sigma_k^{old})} \qquad \ldots(23)$$

for updating the conditional probabilities, and

$$\mu_k^{new} = \frac{\sum_{n=1}^{N} x_n p(k|\overline{x}_n, p^{old})}{R \sum_{n=1}^{N} p(k|\overline{x}_n, pp^{old})}, k = 1, 2, \ldots, K, \qquad \ldots(24)$$

and

$$(\sigma_k^{new})^2 = \frac{\sum_{n=1}^{N}(x_n - \mu_k^{new})^2 p(k|\overline{x}_n, p^{old})}{R\sum_{n=1}^{N} p(k|\overline{x}_n, p^{old})}, k = 1,2,\ldots,K. \quad \ldots(25)$$

for updating the means and variances.

We treated the training set in the data concerning 27 ALL and 11 AML patients as 38 independent repeats of measurements of gene expression and used the iterative formulas (eq.23)(eq.25) to perform a decomposition of the probability density functions into 11 Gaussian components. We excluded genes with expression levels less than zero (erroneous measurements) owing to necessity of taking logarithms. More precisely, we included in the analysis only genes with strictly positive values of their expression levels in all 38 scans. The number of Gaussian components, 11, was chosen arbitrarily. The resulting decomposition, described by estimated values $\hat{\alpha}_1, \hat{\alpha}_2, \ldots, \hat{\alpha}_{11}$, $\hat{\mu}_1, \hat{\mu}_2, \ldots \hat{\mu}_{11}$, and $\hat{\sigma}_1, \hat{\sigma}_2, \ldots, \hat{\sigma}_{11}$, is shown graphically. The upper left plot shows the

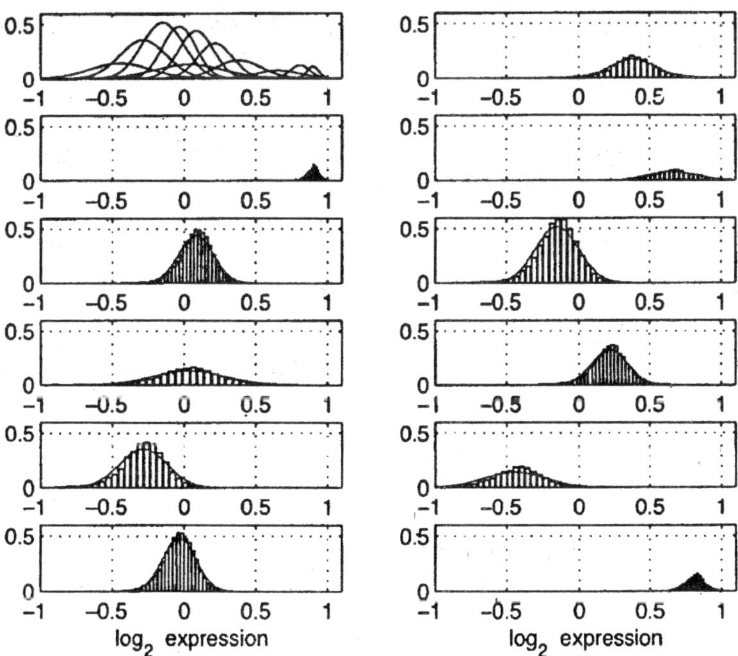

Fig. 2.3. Decomposition of probability density function into 11 Gaussian components, obtained by interpreting the data for 27 ALL and 11 AML patients in as 38 independent repeats of measurements of gene expression and using the iterative formulas.

probability density functions of all 11 components. The plots are scaled by the weights $\hat{\alpha}_1, \hat{\alpha}_2, \ldots, \hat{\alpha}_{11}$, (the areas under curves 1,2,..., 11 are equal to $\hat{\alpha}_1, \hat{\alpha}_2, \ldots, \hat{\alpha}_{11}$, respectively). The other plots show, separately, the probability density functions of the components along with histograms of the logarithms of the expressions of the genes which belong to those components. By "gene n belongs to the kth component" we mean that $k = \arg\max p(x|\bar{x}_n, \hat{p})$, where $p(k|\bar{x}_n, \hat{p})$, is the probability density function given by (eq.23) with $\alpha_1^{old}, \alpha_2^{old}, \ldots, \alpha_K^{old}$, $\mu_1^{old}, \mu_2^{old}, \ldots, \mu_K^{old}$, $\sigma_1^{old}, \sigma_2^{old}, \ldots, \sigma_K^{old}$ replaced by $\hat{\alpha}_1, \hat{\alpha}_2, \ldots, \hat{\alpha}_{11}$, $\hat{\mu}_1, \hat{\mu}_2, \ldots \hat{\mu}_{11}$, $\hat{\sigma}_1, \hat{\sigma}_2, \ldots$, $\hat{\sigma}_{11}$. In other words, gene n belongs to the kth component if $p(x|\bar{x}_n, \hat{p})$ has its maximal value at $x = k$.

Introducing the requirement of that all values for genes should be positive resulted in confining the number of genes analyzed to 2568. Comparing the number of genes in the Hu6800 Affymetrix microarray 6817 with the number of genes included in the analysis, only 2568 we can see that the demand that a gene must give positive expression levels in all scans can be quite tough for some datasets and may lead to the elimination of many measurements. It would be both possible and reasonable to relax our data filter. However, we restricted the analysis to 2568 genes since it was intended mainly to serve for illustration of the methodology. We can understand the 11 normal components as 11 clusters of coexpressing genes; the numbers of genes in these components are 174, 31, 75, 344, 487, 209, 257, 381, 196, 358, and 56. We can see, by the quite good agreement between the histograms of the logarithms of the expression levels of the genes clustered into components and their pdf envelopes, that the model (eq.21) seems to fit the analyzed microarray data quite well.

An issue to be solved in designing computational approaches is how many components K should appear in the sums (eq.18) and (eq.21). When we use a mixture model only for approximation of the probability density function, then the number of components in the mixture distribution is not of great importance and we can afford to apply mixtures with some excess of components. However, when we use mixture decomposition to study the structure of a process and we group genes into different clusters by their different patterns of behavior under different experimental conditions, the problem of deciding on the number of clusters becomes much more important. It can be demonstrated that the quality of clusters and their estimated confidences are rather sensitive to the choice of K. If K is too small, different patterns become clustered together; if, on the other hand, K is too large, the model becomes overparametrized and some components are unnecessary and noninformative. Estimation of the number of components in a mixture distribution model can be done by penalizing over-parametrization with the use of information criteria, such as Bayesian information criterion (BIC). The BIC leads to a penalty for adding parameters (1/2)#parameters*log(#observations). So the BIC-corrected log likelihood functions (eq.18) and (eq.21) will take the form

$$L^{\text{BIC corrected}} = L - \frac{1}{2}\#\text{parameters} * \log(\#\text{observations}) \quad ...(26)$$

where L on the right-hand side is given by either (eq.18) or (eq.21), and #parameters and #observations (or #measurements) are straightforward to determine. We can repeat the maximization of the likelihood (eq.18) or (eq.21) with different values of K (or, better, maximize the likelihood interactively while modifying K) and then find the value of K which maximizes (eq.26). There is evidence that the use of (eq.26) leads to quite reliable estimates of K. A numerically efficient approach to estimating the number of components the use of the Metropolis–Hastings sampling algorithm, penetrates the space of the parameters of the mixture distributions and the space of possible values of K simultaneously. Another possibility is to use using infinite mixture models, where the number of components K is neither limited nor penalized, but assumptions about the prior distributions of the parameters of the Gaussian components are made, analogous to those presented in the maximum likelihood approach.

Class Prediction and Class Discovery

In this section we make some general remarks concerning class prediction (supervised classification) and class discovery (unsupervised classification). These remarks will be developed further and supported by computational examples in later sections.

Two tasks related to the classification of expression profile data are class prediction and class discovery. Class prediction uses information about the expression profiles and the known classification of the data sets or experiments to construct classifiers applicable to future data. When the expression profiles belong to two known classes A and B, a very simple and rather efficient solution is often programmed into microarray scanner software: to pick out genes whose expression is strongly correlated with their class, i.e., underexpressed in A and overexpressed in B, or vice versa. Checking a large number of genes for correlation with a partition of the samples into two (or possibly more) classes touches on the statistical problem of multiple testing.

Other approaches involve linear or nonlinear classification, artificial neural networks, Boolean or Bayesian networks, fuzzy logic classifiers, etc.

Class discovery, i.e., unsupervised classification, concerns telling sets of data apart without using any prior information about the number of classes and/or on the categorization of data. Class discovery is not used for constructing classifiers for future data but, rather, (i) to confirm that the information implied in the design of the experiment, is also encoded in the gene expression profiles collected, and (ii) to explore the data from the angle of existence of unknown relations and mechanisms and to formulate hypotheses explaining these mechanisms. Successful class discovery that is consistent with the prior knowledge about the data and the experiment significantly increases the confidence that inference based on the measured gene expression profiles will prove reliable and robust. The numerical procedure most often applied for class discovery is a hierarchical clustering algorithm which allows one to infer a tree for the arrays and/or genes based on distance matrix. Clusters are then obtained by cutting the branches of the tree at some level. Several variants of this method are possible, depending on the definition of the distance function and on the assumption about how the distance between clusters will follow from the distances between the members of the clusters. Other approaches to class discovery, such as K-means clustering, self organizing maps (SOMs), and Kohonen neural networks. are also used in analyses.

Dimensionality Reduction

A characteristic property of experiments with DNA microarrays is the very large number of genes, which can reach the order of tens of thousands, versus the relatively small numbers of samples (microarrays). This can be an obstacle to extraction of information from the experiments. Techniques of dimensionality reduction by selection of the genes that capture most of the variation in the data are therefore very often applied to DNA microarray data. These techniques include two main approaches: principal component analysis (PCA) and partial least squares analysis (PLS). The computational aspects of these methods involving singular value decomposition of the matrices of measured expressions and formulating the PLS analysis as a sequence of optimization problems. These methods are either combined with class discovery and prediction, or used as a primary source of information about which genes are correlated most with the process analyzed.

PCA aims at identifying major directions in the data space by singular value decomposition of the data matrix and by limiting the analysis to the genes most strongly correlated with the principal directions. PCA is an unsupervised approach; it explores the structure of the data without prior knowledge on experimental conditions, outcomes of experiments, and so forth. The PLS method has the same aim as PCA, to reduce dimensionality, but in contrast to PCA, it needs a measurement or output vector to introduce structure into the data space. In the PLS method, orthogonal directions in the data space are computed on the basis of maximization of the variance between the output vector and a gene expression vectors (or a combination of them). The use of the PLS method requires access to a continuous measurement related to each array. Since this is not always available, the use of the PLS method with microarray data is not reported in the literature on statistics for DNA microarrays as often as the use of PCA. Examples of the application of the PLS method to reduce the dimensionality of DNA microarray, where a survival-time variable was used as an output vector in the PLS algorithm.

Example of Application of PCA to Microarray Data

We shall illustrate dimensionality reduction for DNA microarray data. As previously, we have included in the analysis only genes with strictly positive values of the expression level in all 38 scans, which confined the number of genes analyzed from the 6817 genes in Hu680 to 2568. Taking logarithms of the expression levels, we obtained a matrix X with $m = 2568$ rows and $n = 38$ columns. Each row in the matrix X corresponds to one gene, and each column corresponds to one experiment (a sample from a patient with either ALL or AML). We centered each of the rows (i.e., we subtracted from each element of the row the mean across the row) and applied the SVD (4.18) to the resulting matrix $X^{row\text{-}centered}$. We used the "economy" variant of SVD (4.44), which substantially speeds up computations. The number of nonzero singular values was $r = 37$, one singular value was equal to zero, which is related to the centering of the rows of the matrix X. The top ten singular values capture 65% of the total variance in the data.

Class Discovery

As already said, several different variants of class discovery procedures are applied as steps in data analysis; they are aimed at exploring the data structure and confirming the agreement of the results of unsupervised analysis with the existing knowledge about the data.

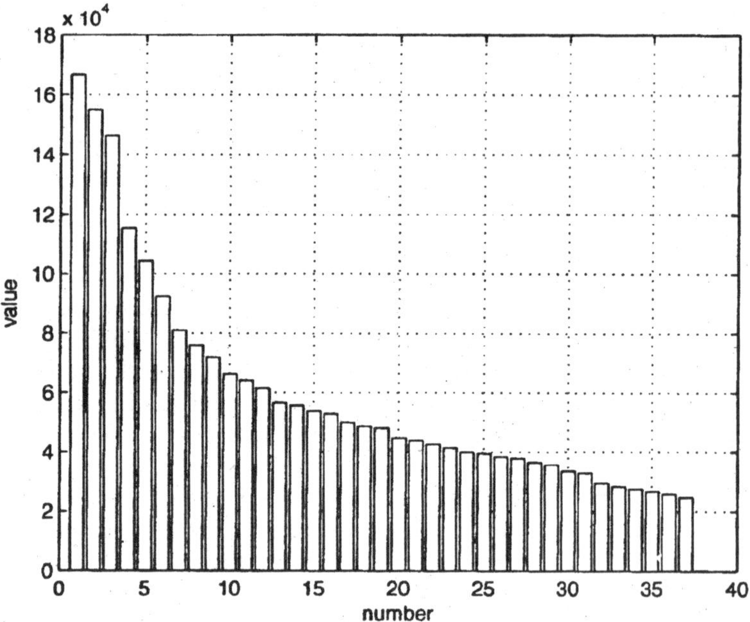

Fig. 2.4. Computed spectrum of singular values.

For example, devoted to gene expression profiles of blood samples of AML and ALL patients, the two-cluster SOM technique was applied to all 38 expression profiles without assuming any knowledge about the data. The results of this unsupervised classification were in good consistency with the prior knowledge on the data. Namely, in the SOM classification only one AML case was misclassified as ALL, two ALLs were misclassified as ALL and for two samples, one ALL and one AML, the classification was not resolved.

The basic interest concerning the data, is whether there are differences of expression profiles between samples, and, more important, whether there are differences in expression profiles between two groups of ALL and AML patients. One must cope with the problems already outlined: the same gene can be marked as present in one patient's sample and absent (missing) in another; or can be measured as positive in one sample and erroneous in another. It may seem desirable to include as much of the data in the computational algorithm as possible, but practical experience shows that researchers must make a compromise between quantity and quality of data. Including genes that are weakly correlated or uncorrelated with the process studied, with many erroneous and noisy measurements, results in the addition of a source of disturbance, which, rather than providing new observations, obscures the relations the research program is after.

Therefore, studies based on gene expression data use steps aimed at excluding genes which add noise rather then contribute with useful information. Some typical approaches are:

1. Exclude genes with multiple erroneous measurements, or labeled as absent in most (or all) of the experiments.
2. Estimate variances, by computing standard deviations, of genes across different measurements with different experimental conditions. Use the hypothesis that the genes with highest variances are the most informative and, in the further analysis, use only genes with high enough standard deviations.
3. Perform principal component analysis and include to further analysis only genes with a high enough affinity to the principal components.

Hierarchical Clustering

In the interpretation of results of experiments involving DNA microarrays, hierarchical clustering is probably the most often used algorithm for class discovery. There are several variants of hierarchical clustering, depending on the definition of the distance and the definition of linkage when clusters are formed. So there is a lot of flexibility in tuning the most convenient version of the algorithm to our needs. Also, with the help of appropriate software, the results of hierarchical clustering can be displayed graphically as trees, with a lot of information presented in a convenient and comprehensive manner.

Here we show an example of nonsupervised data exploration using data preprocessing and hierarchical clustering, again based on the study of ALL versus AML gene expression profiles. We have reanalyzed the data, applying the following steps in the analysis:

1. We included into the analysis genes with strictly positive values of the expression level in all 38 scans, which reduced the number of genes analyzed from the 6817 genes on the Hu680 chip to 2568.
2. The data was log transformed.
3. We assumed the hypothesis that the useful information, concerning classification between ALL and AML, is carried by the genes with the largest variation between samples. We selected 110 genes with the highest values of standard deviations.
4. For the 110 genes selected in step (3), we applied a hierarchical clustering algorithm.

The above procedure is based on the heuristic assumption that genes with more variation are more informative for data classification. The hierarchical clustering in step (4) was performed with the options of Euclidean distance and complete linkage. A classification into two groups can be obtained by cutting the tree at the highest level. One can see that the unsupervised classifier performs quite well, comparably to the class predictor. There is one mistake of classification, the AML sample number 29 is clustered together with ALL samples 127.

Class Prediction: Differentially Expressed Genes

Class prediction involves using information about the classification of the samples under study or about the different experimental conditions for different samples, in conjunction with the gene expression profiles, to classify samples obtained in further experiments. A very simple and rather efficient solution is often programmed into microarray scanner software: to pick out genes whose expression is strongly correlated with classes. Then use these genes as predictors for future classification experiments. The genes whose expressions differ most significantly between different samples or between different experimental conditions are said to be differentially expressed. Lists of these genes not only are of value as predictors or candidates for predictors for the construction of classifiers, but also provide insights into the processes involved in the samples studied, owing to existing knowledge about their functions and interrelations.

For example, many experiments involving the use of DNA microarrays compare gene expression profiles in cancer and normal cells. The comparisons lead to the publication of lists of cancer-versus-normal differentially expressed genes, often ordered with respect to their differentiating power determined by several approaches, for example, by the *p*-value of a statistical test. Genes can then be compared across different studies, and the associated Website containing the collection of microarray datasets of human cancers.

Classifiers based on rather small numbers of the top differentially expressed genes selected in sample comparison studies are usually very effective in predicting taxonomies related to future measurements.

Multiple Testing, and Analysis of False Discovery Rate (FDR)

Creating a list of genes expressed differentially in a sample *A* versus a sample *B* most often involves multiple calls of procedures for testing statistical hypotheses. Since every statistical test accepts or rejects the associated hypothesis with some probability and the testing is repeated many times (typically, the number of tests equals the number of genes), an issue arises of controlling the rate of statistical errors of type I (false discoveries) among the results of the tests. Let us define the family-wide error rate (FWER) as the probability of making at least one (i.e., one or more) type I error (false discovery) in multiple testing.

The idea of control of false discoveries in multiple testing can now be expressed by introducing a constraint on the value of FWER, FWER $< \alpha$. Enforcing FWER $< \alpha$ must be done by introducing corrections concerning the significance levels of individual tests. The corrections, given by the Bonferroni or Sidak formulas or their variants, are known to be very conservative. A more flexible approach is related to introducing a concept of false discovery rate (FDR) equal to the expected proportion of false positives (false discoveries) among the tests where the null hypothesis was rejected (the discoveries).

In a multiple testing procedure based on FDR control was proven to be significantly less conservative than procedures based on the FWER. Further developments were proposed and, in the context of microarray data.

Here we sketch the FDR approach published. This method uses a decomposition of the probability density function related to the distribution of p-values of statistical tests to estimate and control the FDR. Let us assume that a series of DNA microarray experiments have been conducted to compare samples A and B. For each of the genes on the microarray chip, a set of measurements is available and, on the basis of these measurements, a statistical test is performed, with the null hypothesis H_0 being "the distributions $p_A(x)$ and $p_B(x)$ of the expressions of gene X for samples A and B, are equal i.e., $p_A(x) = p_B(x)$". A computer procedure typically included in statistical software packages, returns a p-value of the test which leads either to rejecting H_0 at the significance level α if $p < \alpha$, or to "no premise to reject H_0" if $p > \alpha$ (typically, $\alpha = 0.05$).

Note that the p-value of a statistical test is a random variable with a distribution supported on the segment [0, 1]. Consider two situations:

1. H_0 is true, or in other words $p_A(x) = p_B(x)$. In this situation the p-value of the test is obtained from inverting the cumulative distribution function of the test statistics, so the distribution of p-values is uniform over the segment [0, 1].

2. H_0 is not true, or in other words $p_A(x)$ and $p_B(x)$ are different distributions. If we do not assume a particular form of the distributions $p_A(x)$ and $p_B(x)$ we cannot derive any specific form for the distribution of p-values here. However, in the case when $p_A(x) \neq p_B(x)$, intuitively the p-values of the test should cluster close to zero, since H_0 should most often be rejected. So the distribution is no longer uniform.

In conclusion, the above probability density function, which we denote by $f(p)$, associated with the distribution of p-values of the statistical test applied in (1) and (2), is a mixture of a uniform component $f^u(p)$ corresponding to (1) and a component related to (2) which the authors propose to approximate by a mixture of a number of beta distributions, i.e.,

$$f(p) = w_u f^u(p) + \sum_{j=1}^{J} w_{\beta j} f^{\beta j}(p, a_j, b_j). \qquad \text{...(27)}$$

In the above equation $f^u(p)$ is the probability density function of a uniform distribution; $f^{\beta j}(p, a_j, b_j)$ denotes the probability density function of the jth beta distribution; w_u and $w_{\beta j}$, $j = 1,\ldots, J$ are weighting coefficients; $w_u + \Sigma_{j=1}^{J} w_{\beta j} = 1$, and a_j, b_j, $j = 1,\ldots, J$ are parameters of the beta distributions. There is no direct probabilistic argument for using beta distributions as models for the p-values of statistical tests. However, beta distributions are supported on the unit interval and there is a lot of flexibility in fitting their shapes to data by changing the values of the parameters a and b, which makes them a good tool. The uniform distribution can be seen as a special case of the beta distribution with $a = b = 1$. In most cases it turns out that limiting the approximation to only one beta component, $j = 1$, is satisfactory for approximating $f(p)$ in (eq.27).

In order to write down explicit expression for the false discovery rate, we summarize the possible situations. The numbers $A(\alpha)$, $B(\alpha)$, $C(\alpha)$, and $D(\alpha)$ are random variables; they depend on the threshold chosen α, since the hypothesis H_0 is rejected if $p < \alpha$. Using $A(\alpha)$, $B(\alpha)$, $C(\alpha)$, and $D(\alpha)$ we can compute the false discovery rate as

$$FDR(\alpha) = E\left[\frac{A(\alpha)}{A(\alpha) + B(\alpha)}\right], \qquad \text{...(28)}$$

the ratio of the number cases where H_0 is rejected when it is true to the total number of cases where H_0 is rejected. We also define the discovery rate as

$$DR(\alpha) = E\left[\frac{B(\alpha)}{B(\alpha)+D(\alpha)}\right], \qquad \text{...(29)}$$

the ratio of the number of cases where a false H_0 is rejected to the total number of cases where H_0 is false. Using the decomposition, we can easily compute *FDR* and *DR* in (eq.28) and (eq.29) as

$$FDR(\alpha) = \frac{w_u F^u(\alpha)}{w_u F^u(\alpha) + \sum_{j=1}^{J} w_{\beta j} F^{\beta j}(\alpha, a_j, b_j)} \qquad \text{...(30)}$$

and

$$DR(\alpha) = \frac{\sum_{j=1}^{J} w_{\beta j} F^{\beta j}(\alpha, a_j, b_j)}{\sum_{j=1}^{J} w_{\beta j}}.$$

In the above equations, $F^u(\alpha)$ and $F^{\beta}(\alpha)$ denote the cumulative distribution functions of the uniform and beta distributions.

Table 2.1. Numbers of cases $A(\alpha)$, $B(\alpha)$, $C(\alpha)$ and $D(\alpha)$ corresponding to possible situations associated to accepting or rejecting of the hypothesis H_0.

	H_0 *rejected*	H_0 *not rejected*
H_0 true	$A(\alpha)$	$C(\alpha)$
H_0 false	$B(\alpha)$	$D(\alpha)$

FDR Analysis in ALL versus AML Gene Expression Data

We now give an example of the use of the technique of FDR control again based on the study of ALL versus AML gene expression profiles. As previously, we have included into the analysis only genes with strictly positive values of expression levels in all 38 scans, which reduced the number of genes analyzed to 2568. For all 2568 genes kept in the study we have performed Wilcoxon unpaired sign sum test between the 27 ALL and 11 AML samples. We then fitted the decomposition (eq.27), with one beta component ($J = 1$). The weights obtained were $w_u = 0.54$ and $w_\beta = 0.46$, and a plot of the estimated probability density function is depicted by the solid line. Using the decomposition (eq.27), we computed expected false discovery rates and discovery rates $FDR(\alpha)$ and $DR(\alpha)$, given in (eq.30) and (eq.31).

Gene Ontology Database

The large numbers of genes with different products and functions require bioinformatic tools to make the information about them

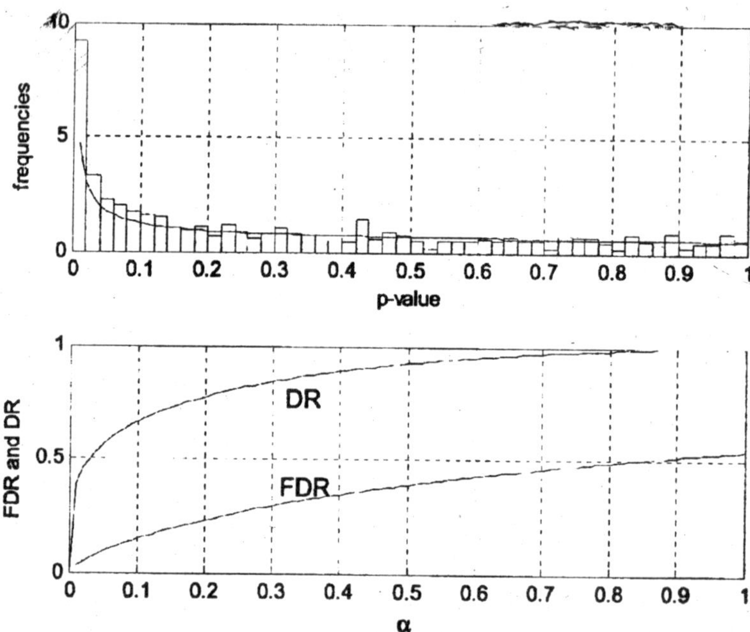

Fig. 2.5. An example of use of the technique of FDR control for the ALL versus AML data.

organized and available. Several databases mentioned in this book store sequences of nucleotides and related sequences of amino acids along with a wealth of background and support information concerning, for example, location, known functions of gene products, and homology between genes in different organisms. Biological studies of sequences and their functions are supported to a great extent by these depositories. For any sequence of nucleotides or amino acids, a researcher can, in just a few minutes, access a large amount of data concerning functions and products of all genes, proteins, and RNA sequences with some kind of similarity to it. Bioinformatic databases have a relational structure, with a dense network of links between items recorded in different depositories.

However, the processes of browsing through genomic and proteomic databases undertaken in many biological research projects have demonstrated the need for more consistent and more structured descriptions of gene products in the various databases. More precisely, many biological studies involving database searches for the purpose of comparison of experimental work performed in the laboratory with references in the literature and data in depositories have experienced inefficiencies and obtained misleading results, because there was not enough structure and precision in the terminology. This situation has been responded to by recent initiatives to create ontology databases. In this section, we focus on the Gene Ontology (GO) database, which is probably the best known and most often referenced ontology database. By gene ontology we mean a standardized and structured vocabulary that describes genes and their products. Ontologies can also involve areas other than gene products.

It may seem that the development of efficient terminology and vocabulary should an inherent feature of progress in science, but in the presence of the massive amount research in the area, the Gene Ontology database maintains the homogeneity of the terminology and makes possible to compare and summarize the results of many studies.

Structure of GO

The terms in GO are organized as a tree structure with three main branches, or rather, GO consists of three ontologies named "*molecular function*", "*biological process*" and "cellular component". "Molecular function" describes the activities of gene products at the molecular level. For example, the terms

in the category of "*molecular function*" include "*nucleotide binding*", "*peptide receptor activity*" or "*lipid binding*". The "biological function" ontology involves naming processes or series of events which can incorporate several molecular mechanisms. Some examples of GO terms classified as "*biological process*" are "*apoptosis*", "*oxygen transport*", "*metabolism*", or "response to stimulus". Finally, the "*cellular component*" GO terms include "*cell nucleus*", "*membrane*", "*ribosome*", and so forth.

The GO database can be explored or downloaded in many formats, including XML, and searched with the use of many different search protocols. There are also numerous programs and Web sites that perform GO searches along with standard queries, for example, BioMart, Babelomics, and AmiGO.

As an example, we can do a very simple query involving the TEL1 yeast gene. By typing in the term "TEL1" and choosing the option "gene symbol/name", we find under TEL1 YEAST the summary

(IDA) — protein kinase activitymolecular function,

(TAS) — phosphorylation-biological process,

(TAS) — response to DNA damage stimulus-biological process,

(TAS) — nucleus-cellular component,

describing GO terms related to the yeast TEL1 gene. The abbreviations indicate, to some extent, the certainty level of the classification, IDA stands for "Inferred from Direct Assay" and TAS means "Traceable Author Statement".

Other Vocabularies of Terms

The development of vocabulary terms in GO described above is being accompanied by similar progress in several related areas. Responding to this development in the organization of terminology, ontology browsers, such as those mentioned above, are including vocabulary terms from new areas in their repertoire of analyses. The new vocabulary terms include genetic and metabolic pathways, protein domains and functional sites, transcription factors, and regulatory elements. The developments in ontology browsers involve both the addition of new vocabularies and the creation of mappings between these vocabularies.

A fast-growing dictionary of terms involves genetic pathways, signaling, metabolic, and regulatory. A large database containing data on the genetic pathways is KEGG, the Kyoto Encyclopedia of Genes and Genomes. KEGG is a Website that organizes databases and associated software, integrating PATHWAY, BRITE, GENES, and LIGAND. Therefore it includes information both on genetic pathways and on related biochemical and biological processes. A common way of using KEGG terms is by analyzing lists of words given by names of pathways corresponding to a set of genes.

Another dictionary of terms concerns interPro motifs related to protein families, domains and functional sites. The related database is run by EBI, European Bioinformatic Institute.

One more example is cisRED database of cis regulatory elements database. The cisRED database holds conserved sequence motifs identified by genome scale motif discovery, similarity, clustering, co-occurrence and co-expression calculations.

Supporting Results of DNA Microarray Analyses with GO and other Vocabulary Terms

The methodologies commonly applied in DNA microarray assays lead to comparisons of large numbers of genes and to grouping of genes into classes or clusters, using a criterion based on correlations or similarities of their patterns of expression. Although many meaningful insights have been obtained with this approach, owing to the large biological variation there is still a lot of uncertainty and ambiguity regarding the meaning of the measured gene expressions. Therefore including GO classes and terms in DNA-microarray-based studies allows automated confronting and combining of the experimental results obtained with the "background knowledge". Since the GO database is actually a tree of names, it can be very easily searched foe single terms as well as for lists of terms, trees, graphs, etc. This makes it a very convenient tool for creating or, rather, adding interpretations to summaries and comparisons of gene expression.

Some of the most obvious approaches to supporting DNA expression analysis by means of GO searches are the following:

1. By using some criterion, obtain a list of genes differentially expressed in two experiments and use the GO terms of the genes in the list to infer possible mechanisms and processes.
2. Obtain two lists of genes, for example, a list A of genes upregulated in experiment A and a list B of genes upregulated in experiment B, and compare these two lists by use of their GO terms.
3. Obtain from GO a list of genes related to, for example, one biological process, apoptosis, and study the pattern of gene expression in some microarray measurements, limited to the list obtained from GO.

3

NUCLEIC ACID DATABASE

The *Nucleic Acid Database* (NDB) was established in 1991 as a resource for specialists in the field of nucleic acid structure. Over the years, the NDB has developed generalized software for processing, archiving, querying, and distributing structural data for nucleic acid-containing structures. The core of the NDB has been its relational database of nucleic acid-containing crystal structures. Recognizing the importance of a standard data representation in building a database, the NDB became an active participant in the macromolecular Crystallographic Information File (mmCIF) project and was the test-bed for this format. With a foundation of well-curated data, the NDB created a searchable relational database of primary and derivative data with very rich query and reporting capabilities. This robust database was unique in that it allowed researchers to do comparative analyses of nucleic acid-containing structures selected from the NDB according to the many attributes stored in the database.

In 1992, the NDB assumed responsibility for processing all nucleic acid crystal structures that were deposited into the *Protein Data Bank* (PDB); it became a direct deposit site for those structures in 1996. In order to meet data-processing requirements, the NDB created the first validation software for nucleic acids. Until 1998, protein–nucleic acid crystal structures deposited into the PDB were post-processed and then incorporated into the NDB.

When the Research Collaboratory for Structural Bioinformatics assumed the management of the PDB in 1998, the tools developed by the NDB were used to process all macro-molecular structures. The NDB continues to provide a high level of information about nucleic acids and serves as a specialty database for its community of researchers.

Table 3.1. The Information Content of the NDB

Primary Experimental Information Stored in the NDB

Structure summary: descriptor; NDB, PDB, and Cambridge Structural Database (CSD) names; coordinate availability; modifications, mismatches, and drug binding

Structural description: sequence; structure type; descriptions about modifications, mismatches, and drugs; description of asymmetric and biological units

Citation: authors, title, journal, volume, pages, year

Crystal data: cell dimensions; space group

Data collection description: radiation source and wavelength; data collection device; temperature; resolution range; total and unique number of reflections

Crystallization description: method; temperature; pH value; solution composition

Refinement information: method; program; number of reflections used for refinement; data cutoff; resolution range; R-factor; refinement of temperature factors and occupancies

Coordinate information: atomic coordinates, occupancies, and temperature factors for asymmetric unit; coordinates for symmetry-related strands; coordinates for unit cell; symmetry-related coordinates; orthogonal or fractional coordinates

Derivative Information Stored in the NDB

Distances: chemical bond lengths; virtual bonds

Torsions: backbone and side chain torsion angles; pseudorotational parameters

Angles: valence bond angles, virtual angles (involving phosphorus atoms)

Base morphology: parameters calculated by different algorithms

Nonbonded contacts

Valence geometry RMS deviations from small molecule standards

Sequence pattern statistics

Information Content of the NDB

Structures available in the NDB include RNA and DNA oligonucleotides with two or more bases either alone or complexed with ligands, natural nucleic acids such as tRNA, and protein–nucleic acid complexes. The archive stores both primary and derived information about the structures. The primary data include: crystallographic coordinate data, structure factors, and information about the experiments used to determine the structures, such as crystallization information, data collection, and refinement statistics.

Derived information, such as valence geometry, torsion angles, and intermolecular contacts, are calculated and stored in the database. Database entries are further annotated to include information about the overall structural features, including conformational classes, special structural features, biological functions, and crystal-packing classifications.

Some features are derived by different algorithms, and it can be difficult to provide the most reliable values. Whenever possible, the NDB has tried to promote standards that allow structure comparison. An outstanding example of this lack of standards were the problems associated with different values for base morphology parameters produced by different programs. These different values in base morphology meant that it was not possible to compare any two structures by using the numbers in the published literature and that it was necessary to recalculate these values for any analysis.

To help resolve this problem, the NDB cosponsored the Tsukuba Workshop on Nucleic Acid Structure and Interactions to which all the key software developers in this field were invited. It was resolved that a single reference frame would be used to calculate these values and an agreement was reached about the definition of that reference frame. All the programs are being amended so that they will produce very similar values for the parameters. The NDB has recalculated these values for all the structures in the repository and will make them available as output from NDB Searches done over the Web.

Data Validation and Processing

The NDB has created a robust data-processing system that produces high-quality data that is readily loaded into a database. The full capability of this system was recently demonstrated by the successful processing of ribosomal subunits, which are very large and complex structures.

Early on, the NDB adopted mmCIF as its data standard. This format has three advantages from the point of view of building a database: (1) the definitions for the data items are based on a comprehensive dictionary of crystallographic terminology and molecular structure description; (2) it is self-defining; and (3) the syntax contains explicit rules that further define the characteristics of the data items, particularly the relationships between data items. The latter feature is important because it allows for rigorous checking of the data.

Structures are deposited via the Web with the AutoDep Input Tool (ADIT) and then annotated using the same tool. ADIT operates on top of the mmCIF dictionary. In the next stage of data processing, a program called MAXIT checks and corrects atom numbering and ordering as well as the correspondence between the SEQRES PDB record and the residue names in the coordinate files. Once these integrity checks are completed, the structures are validated.

NUCheck verifies valence geometry, torsion angles, inter-molecular contacts, and the chiral centers of the sugars and phosphates. The dictionaries used for checking the structures were developed by the NDB Project from analyses of high-resolution, small-molecule structures from the CSD. The torsion angle ranges were derived from an analysis of well-resolved nucleic acid structures. One important outgrowth of these validation projects was the creation of the force constants and restraints that are now in common use for crystallographic refinement of nucleic acid structures. The program SFCheck is used to validate the model against the structure factor data. The R factor and resolution are verified and the residue-based features are examined with this program. Once an entry has been processed satisfactorily, it is released based on its author-defined hold status.

Database and Query Capabilities

The core of the NDB project is a relational database in which all of the primary and derived data items are organized into tables. At present, there are over 90 tables in the NDB, with each table containing 5 to 20 data items. These tables contain both experimental and derived information. Example tables include: the citation table, which contains all the items that are contained in literature references; the cell dimension table, which contains all items related to crystal data; and the refine parameters table, which contains the items that describe the refinement statistics.

Interaction with the database is a two-step process. In the first step, the user defines the selection criteria by combining different database items. As an example, the user could select all B-DNA structures whose resolution is better or equal to 2.0Å, whose R-factor is better than 0.17, and that were determined by the authors Dickerson, Kennard, or Rich. Once the structures that meet the constraint criteria have been selected, reports may be written using a combination of table items. For any set of chosen structures, a large variety of reports may be created. For the example set of structures given above, a crystal data report or a backbone torsion angle report can be easily

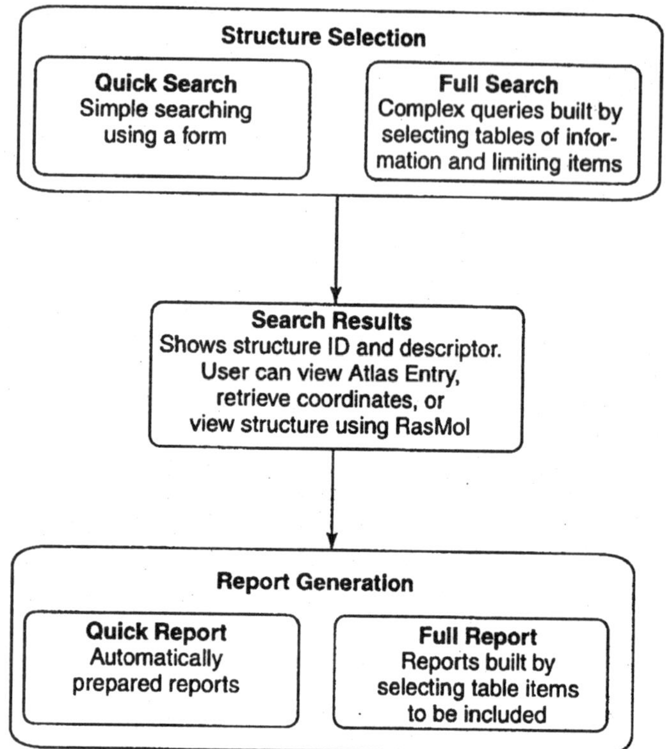

Fig. 3.1. Flow chart demonstrating the two steps involved in searching the NDB: structure selection and report generation.

generated, or the user could write a report that lists the twist values for all CG steps together with statistics, including mean, median, and range of values. The constraints used for the reports do not have to be the same as those used to select the structures.

A Web interface was designed to make the query capabilities of the NDB as widely accessible as possible. In the Quick Search/Quick Report mode, several items, including structure ID, author, classification, and special features, can be limited either by entering text in a box or by selecting an option from the pull-down menu. Any combination of these items may be used to constrain the structure selection. If none are used, the entire database will be selected. After selecting "*Execute Selection*," the user will be presented with a list of structure IDs and descriptors that match the desired conditions. Several viewing options for each structure in this list are possible. These include retrieving the coordinate files in either mmCIF or PDB format, retrieving the coordinates for the biological unit, viewing the structure with RasMol, or viewing an NDB Atlas page.

Preformatted Quick Reports can then be generated for the structures in this result list. The user selects a report from a list of 13 report options, and the report is created automatically. Multiple reports can be easily generated. These reports are particularly convenient for quickly producing reports based on derived features, such as torsion angles and base morphology.

In the Full Search/Full Report mode, it is possible to access most of the tables in the NDB to build more complex queries. Instead of limiting items that are listed on a single page, the user builds a search by selecting the tables and then the items that contain the desired features. These queries can use Boolean and logical operators to make complex queries.

Table 3.2. Quick Reports Available for the NDB

Report Name	*Contains*
NDB Status	Processing status information
Cell Dimensions	Crystallographic cell constants
Primary Citation	Primary bibliographic citations
Structure Identifier	Identifiers, descriptor, coordinate availability
Sequence	Sequence
Nucleic Acid Sequence	Nucleic acid sequence only
Protein Sequence	Protein sequence only
Refinement Information	R-factor, resolution, and number of reflections used in refinement
NA Backbone Torsions (NDB)	Sugar-phosphate backbone torsion angles using NDB residue numbers
NA Backbone Torsions (PDB)	Sugar-phosphate backbone torsion angles using PDB residue numbers
Base Pair Parameters (global)	Global base pair parameters calculated using curves
Base Pair Step Parameters (local)	Local base pair step parameters calculated using curves
Groove Dimensions	Groove dimensions using Stoffer & Lavery definitions from curves

After selecting structures using the Full Search, a variety of reports can be written. The report columns are selected from a variety of database tables, and then the full report is automatically generated.

Multiple reports can be generated for the same group of selected structures; for example, reports on crystallization, base modification, or a combination of these reports can be generated for a particular group of structures.

Data Distribution

Coordinate files, database reports, software programs, and other resources are available via the ftp server. In addition to links to information provided from the ftp server, the Web server provides a variety of methods for querying the NDB. These sites are updated continually.

The NDB Archives, a section of the Web site, contain a large variety of information and tables useful for researchers. Prepared reports about the structure identifiers, citations, cell dimensions, and structure summaries are available and are sorted according to structure type. The dictionaries of standard geometries of nucleic acids as well as parameter files for X-PLOR are available. The NDB Archives section links to the ftp server, providing coordinates for the asymmetric unit and biological units in PDB and mmCIF formats, structure factor files, and coordinates for nucleic acid structures determined by NMR.

A very popular and useful report is the NDB Atlas report page. An atlas page contains summary, crystallographic, and experimental information, a molecular view of the biological unit, and a crystal-packing picture for a particular structure. Atlas pages are created directly from the NDB database. The entries for all structures in the database are organized by structure type in the NDB Atlas.

Mirrors

The NDB is based at Rutgers University and is currently mirrored at three other sites: the Institute of Cancer Research UK, the San Diego Supercomputer Center in San Diego, California, and the Structural Biology Centre in Tsukuba, Japan. These mirror sites are updated daily, are fully synchronous, and contain the ftp directories, the Web site, and the full database.

Community Outreach

The NDB works closely with the research community to ensure that their needs are met. A newsletter is published electronically and provides information about the newest features of the system. To subscribe, send an email to ndbnews@ndbserver.rutgers.edu. Very complex queries will be done by the staff in response to user requests via e-mail.

Applications of the NDB

The NDB has been used to analyze characteristics of nucleic acids alone and complexed with proteins. The ability to select structures according to many criteria has made it possible to create appropriate data sets for study. A few examples are given here.

The conformational characteristics of A-, B- and Z-DNA were examined by using carefully selected examples of well-resolved structures in these classes. Conformation wheels for each conformation as well as scattergrams of selected torsion angles were created. These diagrams can now be used to assess and classify new structures. Studies of B-DNA helices have shown that the base steps have characteristic values that depend on their sequence. Plots of twist versus roll are different for purine–purine, purine–pyrimidine, and pyrimidine–purine steps. This particular analysis has been extended to derive energy parameters for B-DNA sequences.

In a series of systematic studies of the hydration patterns of DNA double helices, it was found that the hydration patterns around the bases are well defined and are local. That is, small changes in the conformation of the backbone do not affect the hydration around the bases. It was also found that there are more diffuse patterns around the phosphate backbone that are dependent on the conformational class of the DNA. These analyses were used to attempt to predict the binding sites of protein side chains on the DNA. In a series of protein–nucleic acid complexes, the hydration sites of the DNA were calculated and then compared with the location of the amino acid side chain. The results were surprisingly good in that in most cases the side-chain site and hydration sites were very close. This result was true even in the case of a very bent DNA that is bound to Catabolite Activator Protein.

Systematic studies of the interface in protein–nucleic acid complexes have been done. In one analysis of protein-DNA complexes, 26 complexes were selected in which the proteins were nonhomologous. The results showed that there are amino acid propensities at the interface that are markedly different than in protein–protein complexes. It was also possible to place the complexes into three classes: double-headed, single-headed, and enveloping. A similar analysis has also been done for protein-RNA complexes. There have also been detailed analyses of the hydrogen-bonding patterns at the protein DNA interface and it was found that CH•••O bonds are surprisingly common. Some analyses have been done on the relationships of crystal packing and conformation.

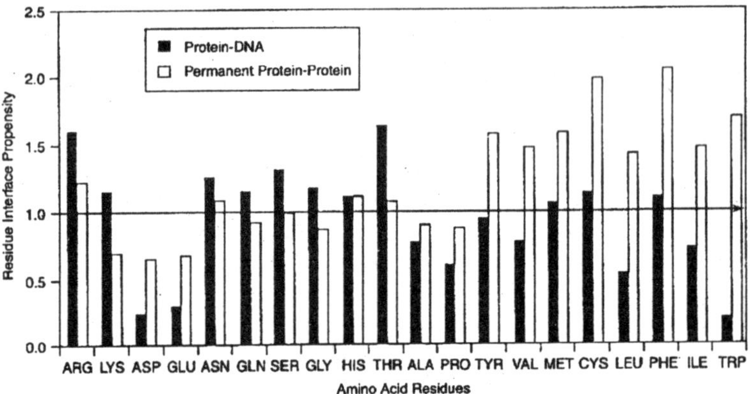

Fig. 3.2. Histogram of the interface residue propensities calculated for 26 protein-DNA complexes and compared to those for permanent protein-protein complex.

Although there are more than 30 different crystal forms of B-DNA in the NDB, the actual number of packing motifs remains relatively small, with the most common motifs being minor groove–minor groove, stacking–lateral backbone, and major groove–backbone.

Minor groove–minor groove interactions in which the guanines of one duplex form hydrogen bonds with the guanines of a neighboring duplex are seen not only in dodecamer structures but in an octamer

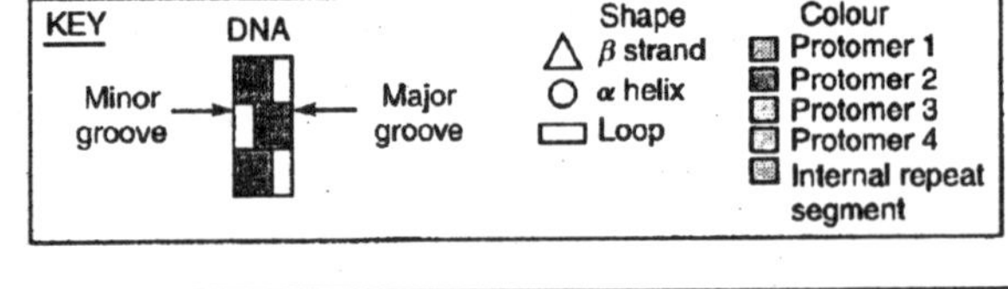

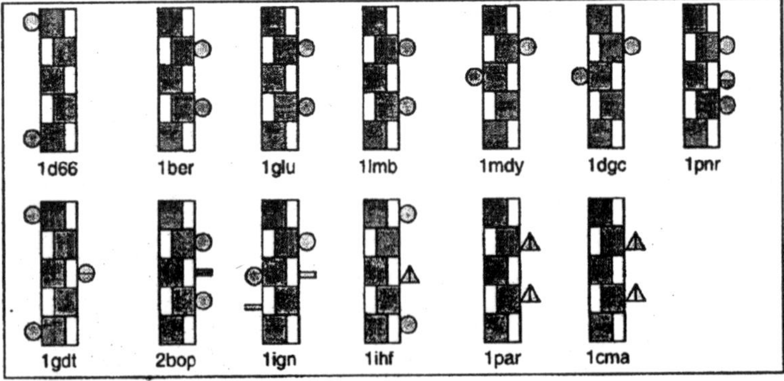

Fig. 3.3. Simple model diagrams of protein-DNA complexes for double-headed binding proteins.

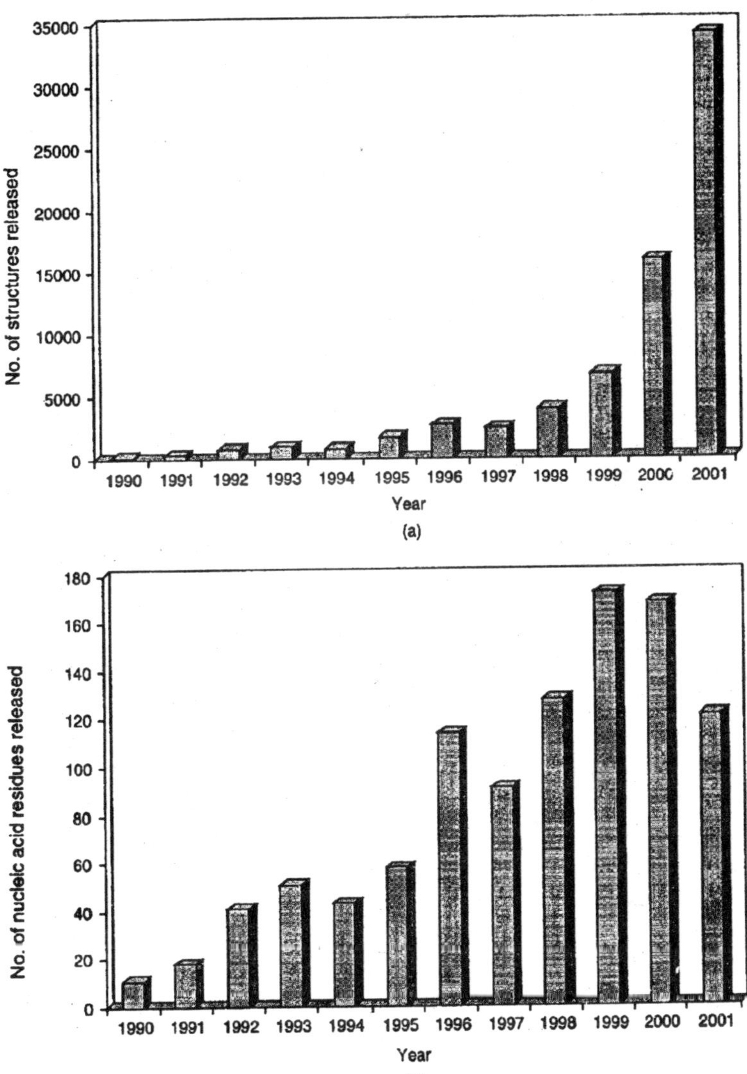

Fig. 3.4. (a) The number of nucleic acid residues, and (b) the number of structures released in the NDB.

sequence with three duplexes in the asymmetric unit. The second motif contains duplexes stacked above one another with the adjoining phosphates forming lateral interactions. A large number of variations of this motif have been observed in decamer and hexamer structures. The third type of packing involves the major groove of one helix interacting with the phosphate backbone of another. Sequence appears to be a large factor in determining these motifs, but it is not the only one. For example, the first structures exhibiting the major groove–phosphate interactions contained a cytosine that formed a hydrogen bond to the phosphate. However, not all structures that show this motif have this hydrogen bond. The particular sequence in this crystal is even more intriguing because it also crystallizes in another crystal form in which the terminal flips out to form a minor groove interaction with another duplex.

The task of trying to determine the relative effects of base sequence and crystal packing on the values of the base morphology parameters is hampered to some degree by the uneven distribution of the 16 different base steps among the different crystal types. Some steps such as CG are very well represented in B structures, whereas others such as AC have very few representatives in the data set. Nonetheless, there are a few steps that occur in crystals with different packing motifs. An analysis of the CG steps across all crystal types shows that its conformation is relatively insensitive to crystal packing and the distribution is similar to that found for all steps. However, the variability of the CA step appears to depend not simply on crystal type but on the packing motif.

The values of twist for CA steps in minor groove–minor groove motifs are smaller than those in the major groove–backbone motif. Very high values are displayed for CA steps in the stacking–lateral backbone motif. Plots of twist versus roll for CG steps show the distribution noted by others and no clustering that depends on crystal type. However, the same plot for CA steps shows very distinctive differences that appear to depend on the packing motifs. It is important to note here that these motifs encompass several crystal types so that the structural variability observed is a function of a particular type of structural interaction rather than a particular crystal form. Before any definitive statements can be made about all the steps it will be necessary to have much more data.

With the recent increase in the number of RNA structures available there have been attempts to establish systems whereby it will be possible to systematically analyze these structures. The result of one of these studies has been the proposal of a classification scheme for the hydrogen bonds in the base pairs (Westhof and Fritsch, 2000). A new syntax (RNAML) has also been proposed for representing RNA structural features.

CHANGING FACE OF THE NDB

When the NDB began, the world of nucleic acid structures consisted of DNA and RNA oligonucleotides, a few protein-DNA complexes, and some tRNA structures. Annotation of structural features was performed manually by visual inspection of molecular architectures. However, since the early 1990s a whole new universe of nucleic acid structures has emerged. There are many ribozyme structures and many different types of protein-nucleic acid complexes represented in over 500 structures. The newest additions to the archive—ribosomal structures—have increased the number of residues of RNA resident in the NDB several fold.

One outcome of the systematic studies that have been done with data from the NDB has been improved classification schemes for understanding nucleic acids. These classification schemes will be used to automatically annotate structures contained within the NDB, which will in turn improve the query capability of the NDB. This type of cycle shows the power of organizing information so that it is more accessible and can ultimately yield new knowledge.

4

Biological Sequences

The information contained in a genome is stored at several levels, the most basic of which associates each amino acid of each protein coded by a gene to a single triplet of DNA bases (codon). Besides this elementary code, simple punctuation signals identify the beginnings and ends of genes. In addition to this 'raw' data, the genome contains expression, regulation, and alternative splicing signals (in eukaryote cells) that govern how cells implement the information it contains. The genome also contains specific signals unrelated to expression of the genetic message, which concern the metabolism of the DNA molecule itself, including replication, recombination, methylation, and restriction sites.

These data are all coded in the DNA sequence, and often mutually overlap. Genes thus contain methylation and recombination sites; certain genes partially overlap; the expression signals of one gene are sometimes located within another.... The unraveling of these various coding levels is of primary importance to the biologist seeking access to information contained in the genome in order to understand the functions of living matter, as well as to devise experimental strategies and to analyze results.

Information technology may be used to efficiently extract the information coded in DNA. The remainder of this chapter recalls and describes various types of signals coded in DNA, as well as specific patterns and sequences with which they are associated.

Genes and the Genetic Code

The principal information contained in the genome consists in the genes themselves. Confronted with the raw sequence data of a genome,

TTT: Phe	TCT: Ser	TAT: Tyr	TGT: Cys
TTC: Phe	TCC: Ser	TAC: Tyr	TGC: Cys
TTA: Leu	TCA: Ser	TAA: **STOP**	TGA: **STOP**
TTG: Leu	TCG: Ser	TAG: **STOP**	TGG: Trp
CTT: Leu	CCT: Pro	CAT: His	CGT: Arg
CTC: Leu	CCC: Pro	CAC: His	CGC: Arg
CTA: Leu	CCA: Pro	CAA: Gln	CGA: Arg
CTG: Leu	CCG: Pro	CAG: Gln	CGG: Arg
ATT: Ile	ACT: Thr	AAT: Asn	AGT: Ser
ATC: Ile	ACC: Thr	AAC: Asn	AGC: Ser
ATA: Ile	ACA: Thr	AAA: Lys	AGA: Arg
ATG: **Met**	ACG: Thr	AAG: Lys	AGG: Arg
GTT: Val	GCT: Ala	GAT: Asp	GGT: Gly
GTC: Val	GCC: Ala	GAC: Asp	GGC: Gly
GTA: Val	GCA: Ala	GAA: Glu	GGA: Gly
GTG: **Val**	GCG: Ala	GAG: Glu	GGG: Gly

Fig. 4.1. The genetic code.

the biologist first seeks to identify the various genes they contain, in order to study the proteins for which they code. Translation of a gene into a protein associates one nucleotide triplet (*codon*) with each of the 20 natural amino acids that constitute proteins. Since there are four different nucleotides, there exist $4^3 = 64$ possible different codon triplets. The meaning of each of these 64 codons is universally conserved throughout all forms of life, and is known as the *genetic code*. Sixty-one codons specify the 20 amino acids, and three, TAG, TAA, and TGA, are translation stop signals, also called *termination* or *nonsense codons*. In addition, protein translation can only be initiated by certain codons known as *start* signals or codons. The main start codon is ATG, which specifies the amino acid methionine; however, translation can also begin with GTG. Whatever the start codon, the first amino acid incorporated into a protein is always methionine, even though the GTG codon normally specifies valine in peptide chains. A gene that codes for a given protein of length *n* therefore has the following structure, known as an *open reading frame* (ORF):

start codon – (n – 1) × {codon specifying an amino acid}
– termination codon

When examining a sequence of cDNA or of a genome that has no introns (bacteria), analyzing the distribution of its nonsense codons furnishes valuable information concerning the potential positions of genes. In a purely random sequence in which the four nucleotides, A, T, G, and C, would be distributed equally, the statistical frequency of occurrence of these stop codons would be 3/64, or around one stop for every 21 codons. However, a protein chain generally consists of between 100 and 1,000 amino acids, which is very significantly longer. The presence of an ORF of length equal to or greater than 100 codons in a DNA sequence is thus a very strong indication of the presence of a gene, thereby providing a predictive method. This simply requires examining stop codon distribution in the six possible reading frames (three in each of the two DNA strands).

Expression Signals

Transcription

The expression of the genes coded in DNA begins with transcription of the genetic message into messenger RNA molecules (mRNA). In eukaryotes, each mRNA molecule codes for only one gene, whereas in bacteria, several genes may be transcribed by the same RNA molecule, constituting what is known as a *transcription unit*. In all cases, the DNA sequence situated around the beginning of the transcription sequence contains a specific pattern known as a *promoter sequence*, which enables the cell to begin polymerizing mRNA.

Cells contain one or several enzyme complexes known as RNA polymerases, which carry out transcription. When transcription begins, certain protein factors or supplementary subunits whose function is to recognize the transcription promoter sequence bind to the RNA polymerase. Once transcription has begun, these factors have fulfilled their roles and detach from the RNA polymerase. For example, in *Escherichia coli*, RNA polymerase is a pentamer, $\alpha_2\beta\beta_2{}'\omega$, to which a supplementary subunit called the σ factor, which recognizes the promoter, is added at the start of transcription. The normal *s* factor, known as σ^{70}, recognizes the 'classical' sequence, located a dozen nucleotides upstream from the beginning of transcription:

$^{5'}$TTGACA- ~17 bases-TATAAT

E. coli cells can also contain *s* factors specific for other DNA sequences that determine other promoter families. Various σ factors are expressed under particular physiological conditions, such as heat shock and low nitrogen levels, in response to which they permit selective activation of the transcription of a whole group of genes.

A comparable, but much more complex situation is observed in eukaryotes, in which three distinct RNA polymerases exist, each interacting with a large number of transcription factors. RNA polymerase II, which transcribes most of the genes for proteins, binds in the vicinity of a conserved pattern known as a *TATA box*, situated around 30 nucleotides upstream from the transcription starting site:

5′TATA (A or T) A (A or T)

This sequence is recognized by a specific protein known as TFIID, by which the polymerase binds. Upstream eukaryote transcription promoters usually include other large patterns that enhance transcription efficiency, such as a 5′-GG(T or C)CAA(T or C)CT around 70 basepairs upstream from the transcription starting site, and sometimes a GC-rich motif of the 5′-GGGCGG or 5′-CCGCCC-type one-hundred basepairs upstream.

Since promoters are found upstream from transcription units, specific downstream sites exist that determine the end of transcription. In prokaryotes, these generally contain an inverted repeat able to fold into RNA, forming a hairpin structure with a stem and a single-strand loop. This sequence is followed by a series of T (U in RNA).

Transcription termination is more complex in eukaryotes. Many eukaryote messenger RNA molecules are polyadenylated at the 3′-end, and the enzyme responsible for this, poly(A) polymerase, recognizes a 5′AAUAAA3′ segment downstream from the gene. This site also appears to be involved in the termination of transcription by RNA polymerase.

Alternative Splicing

The product of DNA transcription in eukaryotes is precursor mRNA, which may contain non-coding segments (*introns*). The precursor RNA then undergoes a process called *alternative splicing*, which excises the introns, yielding mature messenger RNA that can then be translated into protein. This maturation is carried out by complex assemblies of proteins and nucleic acids called small nuclear ribonucleo-proteins (snRNP or SNURPs), which recognize specific sequences at intron-exon junctions:

Exon	Intron	Exon
5′... (C ou A)AG	GURAGU...YYUUYYYYYYNCAG	G...3′

Y indicates a pyrimidine (C or U), R a purine (A or G), and N any nucleotide (A, U, G, or C).

Translation

In eukaryotes, the ribosome sweeps along messenger RNA from the 5′-end, starting translation with the first ATG codon it encounters. In bacteria, where messenger RNA may be polycistronic, the translation start codon is selected by the ribosome, which recognizes the *Shine-*

Dalgarno sequence, AGGAGGU, located 5 to 10 nucleotides upstream from it:

5′AGGAGGU-{5 to 10 nucleotides} (A or G)TG...

Regulation of Gene Expression

A wide variety of mechanisms regulate genetic expression. One or more specialized proteins exist in both eukaryotes and prokaryotes that recognize specific sites on DNA around gene promoter regions. These regulatory proteins can either activate or inhibit transcription, depending on the physiological conditions and external stimuli.

Modification Signals

DNA is not only the medium upon which genetic information is based and that the cell uses to synthesize proteins, it is also a macromolecule that the cell must synthesize, maintain, and repair. DNA therefore also contains signals involved in its own metabolism that are recognized by specialized enzymes of which it is the substrate.

Methylation

Conservation of the genetic heritage is a concern of vital importance for cells, which have developed mechanisms for protecting the quality of their DNA. Polymerization errors may occur during the replication process, resulting in mispairing between the template strand and the complementary strand that is being synthesized. However, cells contain enzymes that can repair such defects. In the repair process, one of the two mispaired bases is excised and replaced by the base that restores correct pairing. The repaired DNA sequence differs according to the stand from which the base is excised:

1. If the excision is on the newly synthesized strand, the repaired double-stranded DNA retains the native sequence;
2. If excision is on the replication template strand, the repaired double-stranded DNA bears a mutation.

The cell has developed a strategy for distinguishing the template strand from the newly synthesized strand. Specialized enzymes introduce chemical modifications (methylation of certain bases) at specific sites on the template strand, but not on the just-synthesized strand, enabling the repair enzyme to distinguish one strand from the other. For example, in *E. coli*, there are two modification systems, *dam* methylase, which methylates adenines in the symmetrical motif

$$^{5'}\text{GATC}^{3'}$$
$$_{3'}\text{CTAG}_{5'}$$

and *dcm* methylase, which methylates cytosines in the quasi-symmetrical pattern

$$^{5'}\text{CC(A or T)GG}^{3'}$$
$$_{3'}\text{GG(T or A)CC}_{5'}$$

Similar specific cytokine methylations exist in animal cells for 5′-CG-3′ sequences and in plants for 5′-CNG-3′ sequences (N can be any nucleotide; A, T, C, or G).

Restriction

An analogous mechanism is used by prokaryotes for defense against bacteriophage viruses. The bacterial cell produces an endonuclease that recognizes a specific DNA sequence and cleaves the two strands of the double helix. This action of this enzyme permits degradation of the viral genome, preventing infection. In order to avoid cutting its own chromosome, the bacteria also contains a methylase that recognizes the same site as the endonuclease and modifies some of the site's bases, inhibiting the nucleolytic activity of the endonuclease. At least one strand of the bacterial chromosome is always methylated at these sites, protecting it from the endonuclease.

Several varieties of restriction enzymes exist, all of which recognize very local, well-defined motifs of between four and a dozen bases. The strands are usually cut at the recognized site, or in the immediate vicinity. Some endonucleases/methylases, such as *Bam*HI, have perfectly symmetrical (*palindromic*) sites:

$$^{5'}\text{GGATCC}^{3'}$$
$$_{3'}\text{CCTAGG}_{5'}$$

Others, such as *Hind*II, have quasi-palindromic sites, with several possible bases at certain positions:

$$^{5'}\text{GT(C or T) (A or G) AC}^{3'}$$
$$_{3'}\text{CA(A or G) (C or T) TG}_{5'}$$

Finally, a small number, such as *Hph*I, have non-palindromic sites:

$$^{5'}\text{TCACC}^{3'}$$
$$_{3'}\text{AGTGG}_{5'}$$

Replication

DNA contains specific patterns involved in replication. Replication usually starts at sites known as *replication origins*, according to various mechanisms that vary as a function of the organism and type of DNA (chromosome, plasmid, or virus). Replication generally begins with the binding of a specialized protein to a nucleotide pattern located at the replication origin. This protein promotes fusion of the double helix, required to initiate replication. For example, in *E. coli*, the chromosome replication origin bears four tandem repetitions of the following nine-base sequence:

$^{5'}$TT(A or T)T(A or C)CA(A or C)A

which is recognized by the initiation protein known as DnaA.

In general, a great number of proteins involved in DNA metabolism 'selective' for certain nucleotide sequences. Therefore it is possible to identify specific patterns at which certain events occur, for example, recombination, integration of a viral genome, and the action of certain gyrases.

Specific Sites

Therefore, as just discussed, identification of the DNA sequence of a site that corresponds to a biological function consists in searching for a pattern. The essential characteristics of the principal types of sites encountered in biological sequences are listed below. Studying these signals reveals the existence of two distinct families, according to whether recognition of the specific motif occurs on template or transcribed RNA. The mechanisms involved are quite different, leading to the identification of two major classes.

Sites Located on DNA

Proteins are Able to 'Read' Double-stranded DNA Bases

As seen in the above examples of signals directly recognized on double-stranded DNA, there is almost always a protein that 'reads' the pattern. To understand how proteins are able to decode a sequence of bases requires a review of the DNA double helical structure.

The two deoxyribose phosphate chains wind around each other on the outside of the molecule, inside which the basepair plates are stacked perpendicular to the axis of the helix. This arrangement forms two lateral grooves of unequal size in which the basepairs are accessible edgewise. Proteins can slide into either of these grooves (most often the larger one), thereby entering into specific interactions with particular basepairs. These basepairs expose different chemical functions, which play either acceptor or donor roles in hydrogen bonding with the lateral chains of the amino acids of proteins lodged in the grooves.

Dimeric Proteins Bind to Palindromic Sites

In analyzing the various types of signals that are recognized by proteins, it becomes clear that a large number of them correspond to palindromic or quasipalindromic sites. For the DNA molecule, this consists of two-fold symmetry around an axis centered on the middle of the motif. In the majority of cases, these sites are recognized by proteins that consist of an even number of subunits that also have two-fold symmetry, dimers, tetramers, etc.... Each half of the DNA site is recognized by one of the halves of the protein, which has the same axis of symmetry as the DNA in the complex.

This symmetrical subunit arrangement can accommodate a large number of regulatory proteins, methylation enzymes, and restriction enzymes, and allows the cell to increase interaction specificity by doubling the size of the pattern recognized without increasing the size of the protein subunit involved in the recognition. This regulatory protein binds to viral DNA on sequences of the following type:

$$^{5'}\underline{\text{TATCACCG}}\text{CCGGTGATA}^{3'}$$
$$_{3'}\text{ATAGTGGCG}\underline{\text{GCCACTAT}}_{5'}$$

and stimulates transcription of genes implicated in the lytic cycle of the virus.

Some Complex Sites Consist of Several Subsites

Certain factors involved in the recognition of specific DNA sites are complex structures, such as nucleoprotein assemblies (for example, ribosomes) and very large proteins consisting of multiple subunits (for example, RNA polymerase). They may therefore be in contact with several distinct regions on the DNA. Each region constitutes an independently recognized sub-site, the spacing between which varies over a more or less elastic range of values. Since the contact zones are discontinuous, therefore distributed over a greater distance, this variable spacing may possibly be due to the flexibility of both the DNA double helix and the molecule with which it is in contact.

Transcription promoters are examples of the most well known complex sites, both in prokaryotes and eukaryotes. They consist of several distinct 'boxes' separated by zones of various sizes. Translation start sites in prokaryotes may also be considered complex sites, since they consist of both the Shine-Dalgarno sequence and the start codon, situated about a dozen bases downstream.

Expression Signals are Often Fuzzy Patterns

Until now, only patterns strictly defined by a given nucleotide sequence have been covered in this chapter, although in some cases, one or several ambiguities have been introduced in the form of alternatives of the type '(A or T)', which has the simple effect of multiplying the acceptable sequences; for example, GT(C or T)(A or G)AC, defines one of the following sequences: GTCAAC, GTCGAC, GTTAAC, and GTTGAC. However, the corresponding sites remain clearly defined and easily recognizable.

Some proteins, such as methylases and restriction enzymes, recognize such strict patterns and tolerate no variation in the admissible sequences. The situation is different with respect to regulation signal sites. In general, it is impossible to define a precise sequence that exactly corresponds to a regulation signal site. However, based on experimental observations, it appears that sets of sequences corresponding to regulation signal expression sites more or less always tend toward an ideal sequence, which is known as a *consensus sequence*.

Such fuzzy patterns allow cells to modulate the efficacy of various expression signals according to their needs. The closer a signal expression sequence approaches to the consensus sequence, the more effective it is. The protein(s) involved can effectively enter into a certain number of molecular interactions (hydrogen and ionic bonding, hydrophobic interactions, etc....) with the bases of the consensus sequence. At positions where the sequence of a site differs from the canonic sequence, interactions with the corresponding bases will be broken. The interaction energy will therefore be lower, which destabilizes formation of the DNA/protein complex to the extent that the site sequence differs from the consensus sequence.

Most of the expression and regulation signal sequences mentioned above are canonic consensus sequence patterns that correspond to maximum effectiveness. For example, the sequences of transcription promoters corresponding to genes that are very strongly expressed in prokaryotes are often practically identical to the following canonical sequence:

$$^{5'}\text{TTGACA} = \sim 17 \text{ bases } -\underline{\text{TATAAT}}$$

Less strongly expressed gene promoters may differ more or less significantly from this ideal sequence, either in the spacing between the two patterns or in the nature of the bases in each conserved cell. In addition, the variation observed among the bases of the consensus sequence is not constant. Constraints on certain positions seem to be greater, probably because the interactions they make with RNA polymerase are stronger and/or more numerous.

This naturally leads to the *weighting* of various bases recognized at a given site. Certain bases may be strictly conserved, and are therefore probably essential, whereas others are less strictly conserved.

Patterns that induce structural changes in the double helix

Until now, the signals discussed have generally constituted sites recognized by other macromolecules, essentially proteins. In addition, specific sequences exist that induce deformations or other structural changes in the DNA molecule, in the absence of an external protein factor. These regions can nevertheless also be the targets of interaction with proteins that then recognize the perturbation in the structure, without necessarily specifically 'reading' its sequence.

The most basic structural change is the appearance of a curvature in the axis of the double helix. The simplest interpretation is that the curvature is caused by an accumulation of deformations at the level of basepair stacking.

The twist angle determines the step of the double helix, which on the average is 34.7° for type B DNA. The roll and tilt angles vary by a few degrees, as a function of the nature of the basepairs, i and i + 1, which can introduce a slight bend. In a given sequence, these slight deformations cancel each other out, resulting in a nearly straight double helix axis. Nevertheless, in particular cases, the small deformations distribute in phase with the step of the helix, and their cumulative effect yields a non-zero curvature.

In particular, the dinucleotide AA (or TT on the complementary strand) causes significant variations in the roll and tilt angles. Some sequences in which series of 3 to 5 A (or T) are regularly spaced every 10 or 11 nucleotides have been described. Since the double helix contains around 10.4 basepairs per turn, all the A in such a region are located on the same side of the double helix; therefore the effects of the induced curve are cumulative.

In general, regularly repeating nucleotide sequence patterns whose periodicity corresponds to that of the double helix indicate the presence of a curved region and/or the binding site for a protein that interacts with one face of the DNA. This sequence is probably curved.

SITES PRESENT ON RNA

Certain Sites are Recognized by other RNA Molecules with which they can Pair

RNA present in the cell is usually single-stranded, which allows it to locally pair with another RNA strand. The sites recognized are usually fuzzy, and the strength of the interaction depends on the stability of the duplex RNA formed. Using the empirical rules to calculate the sites, it is generally possible to estimate their stability. To do this, the sequence of the complementary RNA strand involved in the recognition must be known, which requires studying the underlying biological mechanism.

Among the recognition mechanisms described that utilize RNA: RNA pairing are translation initiation in prokaryotes, in which the 3′ -end of the 16S ribosomal RNA fragment pairs with the Shine-Dalgarno sequence of messenger RNA, and alternative splicing in eukaryotes, in which intron-extron junctions interact with small RNA fragments in the nucleus (snRNP or SNURPs).

Certain sites correspond to local stem and loop RNA folding. Single-stranded RNA can also pair with itself (internal pairing) and fold locally, forming stem and loop structures. Such secondary structures can play multiple roles; for example, they can serve as specific sites for proteins that recognize the shape rather than the sequence of an RNA segment. Inversely, RNA regions involved in internal pairing cannot interact with other RNA molecules. Some sequences may be 'masked' by the formation of a stem and loop structure. Finally, as in the case of DNA transcription terminators, the presence of such a secondary structure can hinder the progress of a polymerization enzyme, leading to a pause or block.

The existence of an RNA segment able to fold into stems and loops is revealed by the presence of extended Palindromic regions (repeated inverse regions), such as the DNA binding sites of dimeric proteins. When such palindromes are detected in a sequence, the biological context must be analyzed in order to distinguish between these two alternatives: a symmetric site on the DNA molecule, and a stem-and-loop structure on the transcribed RNA.

Pattern Detection Methods

The sequencing programs used today generate huge quantities of raw data from which biologically pertinent information must be extracted, such as the locations of genes, expression signals, etc. This analysis essentially involves two complementary approaches:

1. Pattern detection, which consists in locating defined sequence segments that correspond to documented biological functions, such as those described in the preceding paragraphs;
2. Content search, which analyzes the properties of the statistical distribution of nucleotides or amino acids in a sequence.

The following is an intentionally simplified description of the methods used in pattern detection. Known as *pattern matching* in basic computer science, this heavily studied domain employs high-performance algorithms that employ very formal methods. In what follows, these algorithms will be approached using only practical 'biological' examples. For a fully formal treatment of this subject, the reader is advised to refer to specialized works.

Simple Searches

Numerous patterns associated with biological functions are relatively simple in nature. For example, the eukaryotic messenger RNA polyadenylation signal is a non-degenerate hexanucleotide whose sequence is AAUAAA (AATAAA in the corresponding DNA sequence). The search for occurrences of such a non-degenerate pattern in a long DNA sequence may be conducted using the following naïve algorithm:

```
i ← 1
k ← 1
Do
  If Pattern[k] = Sequence[1+k] then
    k ← k+1
  Else
    k ← 1; i ← i+1
  If k > Length(pattern) then
    pattern found at position i
    k ← 1; i ← i+1
While i ≤ Length (Sequence)
```

This simple algorithm consists in nucleotide-by-nucleotide comparison of the pattern with the target sequence at a given position. If the comparison fails, the pattern is advanced one position and the operation repeated. This algorithm has the advantage of being simple; its cost is simply a function of the target sequence length. Nevertheless, it is not optimal, since it may require a number of comparisons (line shaded in gray) that are greater than the length of the sequence itself. To demonstrate this, it suffices to observe what happens when the target sequence is very rich in A and U[1]: For example,

... AAAUAUUAAAUAUUAGACGAAUAAAAGUAUAUUUAG...

At the beginning of the search, the situation is the following:

target: AAAUAUUAAAUAUUAGACGAAUAAAAGUAUAUUUAG

pattern: AAUAAA

test ++-

The naive algorithm will produce two positive comparisons (i) for the two first pairs of A, before failing in the third (-). It will then shift the pattern one position and repeat the analysis. This time, it will fail in the fifth comparison, after four positive tests:

target: AAAUAUUAAAUAUUAGACGAAUAAAAGUAUAUUUAG

pattern: AAUAAA

test +++++-

The algorithm compared a total of eight positions just for the two positions tested. This example was particularly unfavorable, but for strongly biased target sequences, it is possible to conduct twice as many comparisons as the number of nucleotides they include.

This method is not optimal, since when the naive algorithm yields several coincidences between the pattern and the sequence before failing it returns to the target sequence, in which case it reads some target sequence nucleotides several times. It is possible to improve the naïve algorithm by making it 'remember' what has already been read, thus avoiding going back.

Searching with a Finite State Automaton

To keep track of information concerning nucleotides that have already been read, a virtual machine known as a *finite state automaton* is devised. The machine is characterized by a finite number of certain states.

The automaton begins to function in a particular state, called the *initial state*. The machine reads one nucleotide in the sequence during each cycle, then evolves toward a new state, according to the nucleotide and its present state. If the automaton reaches another specific state, known as the final state, the pattern has been found and the automaton stops.

The following three elements constitute a finite automaton:

1. An alphabet Σ of characters for the target sequence (for biological applications, four nucleotides or twenty amino acids);
2. A list E of n states e_0, e_1 ... e_n. This list includes an initial state, e_0, and one or more final states.
3. A transition function T which is an application of E × Σ – E. This function determines the $i + 1$ state, starting from the current state i and the character read. This transition function is stored in the form of a matrix of dimensions (n, k) where n is the number of states and k is the size of the alphabet Σ (four nucleotides or twenty amino acids).

The automaton can then be simulated by applying a very simple algorithm:

```
i ← 1
state ← e0
Do
  State - T[state, sequence[i]]; i ← i+1
If state = finalstate then
  Pattern found at position i; Quit
While i ≤ Length(sequence)
```

This algorithm must read each nucleotide or amino acid in the sequence only once and not go back over it. The problem is obviously to specify the list of states and transition table, so that the automaton will search for the pattern. Taking the example of the eukaryote polyadenylation pattern, AAUAAA, it is possible to achieve this with the automaton.

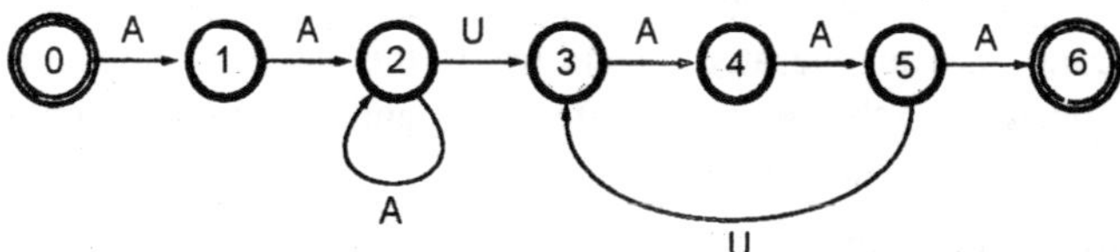

Fig. 4.2. Graphic representation of a finite state automaton seeking a polyadenylation signal, AAUAAA.

This helpful graphic representation permits better understanding of the operating principle of the automaton. In particular, it is possible to see how reading the pattern sought, AAUAAA, leads directly to the final state (state 6). The transition table, T, that corresponds to this graph is indicated below (there is no entry for state 6, since it is the final state). Using the above automaton to analyze the target sequence, the reader may verify whether it goes through the states indicated below and identifies the pattern.

The state of the automaton at a given instant constitutes its 'memory'. It permits knowing which left sub-part of the pattern has already been identified. If the automaton is in state 2, this indicates that it has just read two A (adenine). If it reads a third one, which does not correspond to the sequence sought, AAUAAA, the two last A of the three read still correspond to the beginning of the pattern.

Degenerate Patterns

The Knuth-Morris-Pratt (KMP) method functions well for simple, non- degenerate patterns and may readily be adapted to other, partially degenerate types of patterns, such as consensus sequence sites that give rise to alternative splicing in eukaryotes (exon-intron junction),

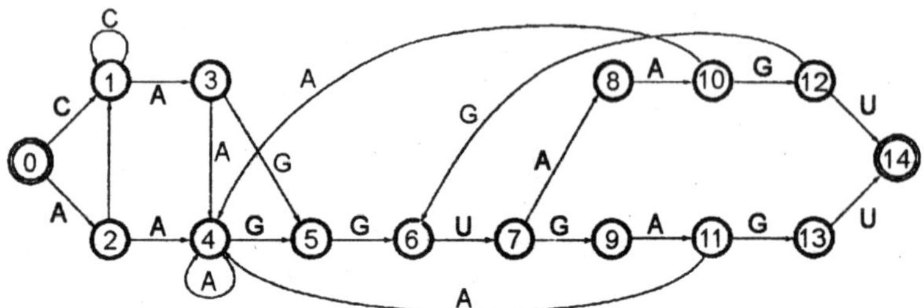

Fig. 4.3. Simplified representation of a finite state automaton that searches for the pattern (C or A)AGGU(A or G) AGU, which corresponds to sites that give rise to alternative splicing in eukaryote RNA.

for example, (C or A) AGGU(AA or C)AGU. Generalizing the KMP approach, it is possible to devise automata capable of seeking such patterns. It is more complex than the one in the preceding paragraph, consisting of 13 states and presenting a branched structure that allows managing the degenerated positions of the pattern. Some arrows corresponding to the return to states 0, 1, and 2 have been omitted in order to simplify the figure.

Regular Expressions

Finite state automata are extremely powerful tools that can recognize a wide variety of different patterns, not only in the context of biological sequences, but in other domains as well. They are often utilized by text and program editing software that includes sophisticated functions of the 'search/replace' type. In general, these search functions employ a language for identifying somewhat complex patterns in the files analyzed. In general, this rather extensive language may be reduced to what are called '*regular expressions*'.

Regular expressions are defined with respect to three basic operations:

1. *Concatenation*, the consecutive linking of two patterns: pattern_1 pattern_2
2. *Alternation* between two patterns, notated 'or': pattern_1 or pattern_2
3. *Repetition* a given number of times (0 or more) of a pattern, notated '*': pattern*

According to the application concerned - word-processing, biological sequences, program editors, etc - the basic language can adopt slightly different syntaxes and take on operators that simplify notation, but which can always be reduced to the above three fundamental operations.

The formal framework of regular expressions allows defining a very wide variety of interesting biological patterns, such as codons, open reading frames, elementary signals, and consensus signals, of which some examples follow. In order to reduce the complexity of the expressions, simple notations are introduced, some of which derive from the syntax of the Unix-based *grep* program.

For the simple alternatives using alphabetic symbols, the following notation within brackets is used:

[ATG] ↔ (A or T or G)

[AG] ↔ (A or G)

Sometimes 'N' or 'X' are used as wild cards to indicate any nucleotide or amino acid, respectively:

N ↔ [ATGC]

X ↔ ACDEFGHIKLMNPQURSTVWY

Among these simple constructions, it is possible to define 'punctuation' codons, which identify the beginning and end of an open reading frame:

start codon: [AG] TG

stop codon: [T([GA]) or A[AG])

A coding codon, that is, a non-stop codon, is longer, but still accessible:

coding codon: [ACG]NN or T([CT]N or G[CGT] or A[CT])

Starting from these constructions, it is possible to define an open reading frame as being a start codon assembly followed by some number of coding codons (identified by a '*' symbol) and ending with a stop codon.

open reading frame: [AG]TG([ACG]NN or T([CT]N or G[CGT] or A[CT])* T([GA] or A[AG])

It is also possible to define variable spacing between patterns, like the one separating the prokaryote ribosome binding site from its start codon (between 6 and 12 nucleotides).

Ribosome binding site: AGGA or GGAG or GAGG

Translation start site: (AGGA or GGAG or GAGG)(N or NN or NNN or NNNN or NNNNN or NNNNNN or NNNNNNN) [AG]TG

This type of pattern definition by regular expression can be very readily generalized to protein sequences, using the 20-character amino acid alphabet instead of the four nucleotides.

The PROSITE pattern library employs regular expression syntax in this way to define more than a thousand patterns and protein signatures.

Finite Automata and Regular Expressions

The fundamental property of patterns defined by regular expressions is that they exactly correspond to patterns that may be sought using finite state automata. In other terms, if it is possible to represent a biological pattern of interest using a regular expression, then it is possible to construct a finite state automaton which seeks it in a sequence, and reciprocally.

The following is an example of an automaton that seeks a pattern which is a simplified version of the expression that defines an open reading frame:

[AG]TG (NNN)* T([GA] or A[AG])

This automaton is different from those described above in that its behavior is not deterministic. For example, if a T is read in state 6 of the sequence analyzed, there exist two possible actions: either going on to state 7, which amounts to trying to identify a stop codon, or returning to state 4; that is, continuing the open reading frame. Therefore the automaton has several different possible routes and *a priori* it is impossible to say which one leads to the final state, state 10, which is why this construction is known as 'a non-deterministic finite state automaton'. If in reading a DNA sequence, one of the possible paths leads to the final state (state 10), then one occurrence of the pattern sought has been found; that is, an open reading frame.

Non-deterministic automata may be used to conduct searches for complex patterns contained in biological sequences. In order to simulate the multiplicity of possible paths in graphs for a non-deterministic automaton, a *set* of states replaces the single state used in deterministic automata. The transition function T must then be modified. Deterministic automata use the application $E \times \Sigma \rightarrow E$, in which E

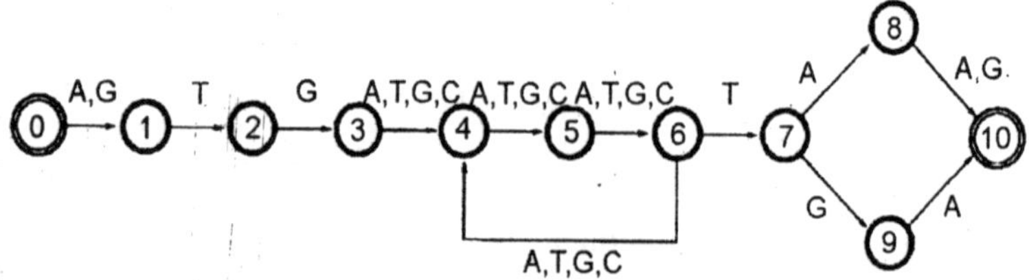

Fig. 4.4. A non-deterministic finite state automaton that recognizes open reading frames. The arrows indicating the return to the initial state are not represented.

is the list of states and Σ the alphabet utilized. For non-determinist automata, starting from a given state e_i, for a character read in Σ, it is possible to end up at several different states. In the example, starting from state 6, when a *T* is read, the automaton ends up in states 7 and 4. The transition function *T* is therefore an application of $E \times \Sigma \rightarrow P(E)$, where *P*(*E*) represents the set of all subsets of *E*

The functioning of such non-deterministic automata may then be simulated by applying the following algorithm, which is a generalization of the one given above for deterministic automata:

i ← 1
state_list ← {e0}
Do
list ← {}
For ei *in* state_list
list ← *union* (T[ei, sequence[I]], list)
state_list ← list
i ← i+1
If final_state ∈ statelist *then* pattern found; *quit*
While i ≤ Length (sequence)

The following is an example of an analysis obtained using this automaton. The sequence read is indicated above the list of states through which the automaton passes upon analyzing each nucleotide. Start and stop codons in the miniature open reading frame detected by the automaton are underlined in the sequence.

Non-deterministic automata may be constructed directly from regular expressions by applying the three basic rules defined above, concatenation, alternation, and repetition. Figure 4.5 is a schematic representation of how to construct automata that recognize the expressions (AB), (A or B), and (A*), starting from automata that recognize the expressions A and B, respectively. The principle of these constructions consists in appropriately merging the initial and/or final states of elementary automata. By iterating these assemblies, it is possible to progressively construct a non-deterministic finite state automaton that corresponds to any regular expression.

Statistics and Sequences

Nucleotide sequences extracted from a database usually seem quite unreadable, and at first sight may be difficult to distinguish from a random series of letters A, T, G, and C. However, this impression is erroneous, since 'real' nucleotide sequences are replete with

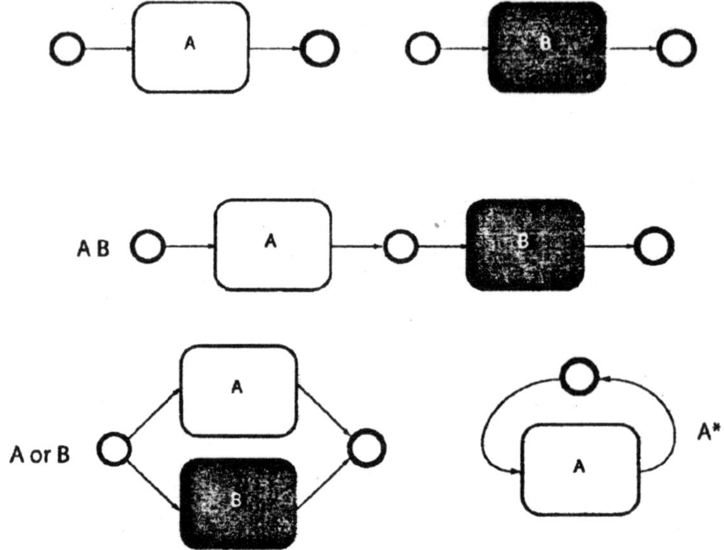

Fig. 4.5. Principles of automaton assembly. The initial and final states are represented by boldface or double circles, respectively.

redundancies and statistical biases resulting from the processes of evolution and natural selection to which they have been subjected. Studying these biases is extremely informative for the biologist, since they provide information concerning the origins of the phenomena responsible for them, leading to a better understanding of how living cells exploit their genetic information. Once the mechanisms involved in such biases have been characterized, their analysis and systematic investigation become valuable tools for use in predicting the properties of other biological sequences.

Nucleotide Base and Amino Acid Distribution

This chapter covers the distribution of nucleotide bases in genomes and of amino acids in proteins. Analysis will proceed in the direction of increasing complexity, beginning with monomer frequency, followed by more complex patterns of length n (n-tuples), including their frequencies and correlations with monomer frequency, as well as with various types of sequences (coding and non-coding strands, introns, exons, etc).

Genome composition

Since the sequence of bases constituting the genome of living species consists of the four nucleotide bases, A, T, G, and C, the first point to consider with respect to genome composition is the relative frequencies of the bases. The principle of base complementarity in the DNA double-helical structure imposes overall equality in the numbers of A and T, and G, and C. What is generally studied is the (G+C)/(A+T) ratio or the GC percent in a given genome or part thereof.

Well before sequencing methods had been developed, physicochemical techniques, such as the measurement of DNA density and fusion temperature, were used to demonstrate the wide disparities in the genomic GC content of various species. These differences were later confirmed by statistical analysis of sequences extracted from databases.

GC/AT distribution in the *E. coli* genome is more or less equal; around 51 to 49 percent. However, equality is not the general rule, and the GC content can range between 15 and 70 percent, according to the species. Thus, only around 18 percent GC is found in the genome of the malaria protozoan *Plasmodium falciparum*, versus 68 percent in *Thermus thermophilus*, a bacterium that lives in hot springs able to thrive in temperatures exceeding 80°C. The overall GC level in vertebrate genomes is intermediate, ranging between 40 and 45 percent. However, the situation is somewhat more complex, since the vertebrate genome is segmented into large regions called '*isochores*', in which the GC level is locally constant, but different from that in neighboring regions.

The origin of the variation in the GC rate among various living species is not entirely clear. In thermophilic organisms, the genome generally includes a high proportion of GC, which permits DNA to

***Plasmodium falciparum* (18 % GC)**

TGTTCACAAACTTCAAACGTGAAAGATTTTAAAACGAGGAGAAGTAAATTTACGA<u>ATG</u>GA
AATAATTATGGGAAAAGTGGTAATAATAAAAATTCGGAGGACCTAGCGATAAACCCAGGT
ATGTATGTTAGAAAAAAAAGAAGGGAAGATAGGGGAGTCCTATTTCAAAGTTCGCAAATTA
GGTAGCGGTGCGTACGGAGAAGTATTATTGTGTAGAGAAAAACATGGACATGGTGAGAAG

***Thermus thermophilus* (68 % GC)**

CGGCCGCGAGCTCCATTTTTTTGACTGCCCTGGAGCCACCCAGGGCGGCTTTGGCATGTA
AGGAGGACGGAG<u>ATG</u>GCGAAGGGCGAGTTTATTCGGACGAAGCCTCACGTGAACGTGGGG
ACGATTGGGCACGTGGACCACGGGAAGACGACGCTGACGGCGGCGTTGACGTTTGTGACG
GCGGCGGAGAATCCGAATGTAGAGGTTAAGGACTACGGGGACATTGACAAGGCGCCGGAG

Fig. 4.6. Sequence of two gene fragments derived from organisms whose genomes contain very different proportions of GC. Initiation codons are underlined.

better resist thermal denaturation, since G-C pairing is more stable than that of A-T.

Comparison of proteins

In contrast to DNA, the average protein composition remains relatively constant throughout all living organisms. However, there is some variations, according to the type of protein considered. For example, membrane proteins are rich in hydrophobic residues, and proteins that interact with nucleic acids are often more basic, but these are general tendencies that depend little on the species. In the 'standard' protein composition, it may be seen that the distribution of the 20 amino acids is unequal: some amino acids are relatively abundant, while others are less common. The average amino acid distribution is a kind of universal signature that may be used to distinguish a real protein sequence from a random one.

N-tuple frequency

Analysis of the incidence of simple nucleotides may be extended to dinucleotides, trinucleotides... and more generally, to *n*-tuples. Triplets will be specifically treated below, in connection with the genetic code.

In this type of analysis, the frequency of occurrence of an *n*-tuple of the bases $B_1B_2 \ldots B_n$, $f_{B1B2} \ldots B_n$, is compared with the product of the frequencies of occurrence of the individual bases $f_{B1} f_{B2} \ldots f_{Bn}$. If it is lower, the *n*-tuple is underrepresented, if higher, it is overrepresented.

This kind of analysis may be carried out on an entire genome, or on two families of disjoined sequences in parallel, in attempting to discriminate between them (e.g., coding or non-coding sequence; intron or exon). Among the most spectacular results of such analysis are:

1. Marked under-representation of the CG dinucleotide throughout all vertebrate genomes. The CG sequence is a cytosine methylation signal. The 5-methylcytosines located at these sites can be transformed into T (thymine) by deamination, a common type of chemical damage. It is clear then that, little by little, owing to the effect of successive mutations, CG sequences can become TG sequences, thus becoming increasingly rare in the genome.
2. Underrepresentation of the CTAG palindromic quadruplet in the *E. coli* genome:

$$f_{CTAG} = 3.6 \times 10^{-4} \ll f_C f_T f_A f_G \approx 3.9 \times 10^{-3}$$

In general, palindromes appear to be underrepresented in bacterial genomes.

Frequency of base triplets; use of the genetic code

The fact that DNA includes genes that code for proteins imposes additional constraints on the nucleotide sequence that constitutes a gene. Since translation of DNA into protein uses elementary words called *codons*, which consist of three bases, it is reasonable to examine the frequencies of the 64 possible triplets. The presence of long open reading frames without stop codons is already a first indication of the non-random character of their distribution.

Respecting the reading frames, it is possible to analyze the frequency of the 61 base triplets corresponding to the 20 amino acids. Codon distribution must necessarily follow that of the corresponding amino acids in the genetic code. For example, on average, tryptophan residues constitute 1.3 percent of proteins; therefore genes should consist of 1.3% TCG, the only codon that corresponds to that amino acid.

It is more interesting to look at what happens in the case of amino acids that are coded by several *synonymous* codons. For example, lysine, which represents 5.7 percent of the amino acids present in proteins, can be specified by two codons, AAA and AAG. Are these two triplets represented in an equal manner? For 100 lysine codons in the human genome, there are only around 38 AAA for every 62 AAG. In the *E. coli* genome, the proportion is reversed, with 60 AAA for every 40 AAG. Cells therefore express a preference for certain synonymous codons. This preference is also species-specific.

It is possible to compute the statistics of the occurrence of various codons in the genes of a species, compiling what is known as a *codon usage table*. The various codon frequencies observed are the result of two superimposed effects: the amino acid composition of the proteins (which is not uniform) and the systematic preference for certain codons among the various possible synonymous codons. Certain triplets, such as the arginine codon AGG in *E. coli* and the alanine codon GCG in the human, which appear to be systematically avoided, are known as '*rare codons*'.

When compiling codon usage tables for different genes derived from the same organism, it is clear that they are all very similar, in general, faithfully reflecting the overall table. Thus, every gene generally conforms to these preference rules, which are a kind of 'signature' of the genome in which it occurs.

In addition, notice that the preferential use of certain synonymous codons is conserved over the course of evolution. The tables of neighboring species often reveal very similar codon usage, and to some extent, their comparison may be used to evaluate their degree of evolutionary relationship.

It has been observed that extreme (very high or very low) GC levels in the genomes of organisms have a significant effect on codon selection. In genomes that are GC-rich, codons ending in C or G are very strongly preferred, whereas the opposite is true for genomes that are AT-rich. Generally, even in species whose genomes have a nearly equal GC/AT ratio, the constraints imposed by the codon usage table result in a more accentuated bias for bases located in the third codon position.

Context-dependent codon biases

In cases in which several synonymous codons are able to code for the same amino acid, each organism expresses preferences that may be expressed as frequencies and listed in a codon usage table. Other effects may add to this scheme of preferences. When several codons of nearly equal probability are possible, the choice among them may be influenced by the neighboring codon; that is, by nucleotides located *immediately* upstream and/or downstream. In *E. coli*, it has been noticed that for the two lysine codons AAA and AAG, the former is more frequently encountered when the next codon starts with a G, and reciprocally, that AAG is preferred when there is a C immediately downstream.

These biases are collectively known as 'context effects,' some of the most marked of which are indicated below for the *Escherichia coli* genome:

AAA-G > AAG-G	AAG-C > AAA-C	Lysine
GAA-G > GAG-G	GAG-C > GAA-C	Glutamate
GGC-G > (GGG-G, GGA-G, GGT-G)	GGT-A > (GGT-C, GGT-T, GGT-G)	Glycine
TTT-G > TTC-G		Phenylalanine

Biological Basis of Codon Bias

Codon utilization adapts to transfer RNA concentrations

It has been shown that the frequency of utilization of each codon in the yeast *Saccharomyces cerevisiae* and in the bacterium *E. coli* is directly proportional to the intracellular concentration of transfer RNA (tRNA) that decodes it. For example, *E. coli* has two transfer RNA isoacceptors for isoleucine, $tRNA^{Ile}_1$, which decodes the ATT and ATC codons (respective frequencies: 2.7 and 2.8 percent), and $tRNA^{Ile}_2$, which decodes the ATA codon (frequency: 0.4 percent). The intracellular

$tRNA^{Ile_1}/tRNA^{Ile_2}$ ratio is 20/1, which is close to the ratio of the decoded codons (27 + 28)/4.

This adaptation optimizes the protein translation process by exactly adjusting the tRNA demand of the translation machinery to the amount available in the cell.

Most highly expressed genes are the fittest

Comparing the codon usage table of individual genes with the '*canonic*' table based on the full genome reveals that the genes which most closely follow the canonic table are those most strongly expressed in the cell (ribosomal proteins, translation factors, structural proteins, etc). The selection pressure on these genes is the greatest, resulting in stronger bias than for weakly expressed genes.

At each stage of the translation process, tRNA molecules in the cytoplasm diffuse into the active site of the ribosome, which 'tests' whether the codonantocodon interaction is correct and 'rejects' inappropriate tRNA molecules. The ribosome 'waits' for the tRNA molecule that possesses the anticodon corresponding to the codon. This wait is proportional to the number of unfruitful tests that the ribosome must carry out before the appropriate tRNA molecule arrives, which depends on the relative abundance of the isoacceptor sought among the set of tRNA molecules:

<number of tRNA molecules tried>

= [total tRNA molecules]/]tRNA molecules sought]

The most abundant glycine codons in *E. coli* are GGT and GGC. They are translated by $tRNA^{Gly_3}$, which accounts for 6.5% of all tRNA in the cell. Ribosomes must therefore carry out around 15 tRNA trials (1/0.065) when translating one of those codons. It takes the ribosome more than 300 attempts to translate a rare codon such as ATA, which codes for isoleucine, since the corresponding isoacceptor represents only 0.3% of the total tRNA. When ribosomes encounter a rare codon, there is a pause in translation. By utilizing only codons that correspond to the most abundant isoacceptors, translation of the most highly expressed genes is able to avoid delays that can slow cell growth.

Rarity of the CG dinucleotide in vertebrates affects the codon usage table

As described earlier, due to the effect of mutations, the CG dinucleotide, which is a methylation site, tends to disappear in vertebrates. As a result, codons that include the CG doublet are underrepresented in vertebrate codon usage tables. This phenomenon is particularly evident for proline (CCN), threonine, (CAN), and alanine (GCN), in all of which a G in the third position is strongly disfavored. This is also reflected in context effects; codons ending in C upstream from a codon that begins with G are systematically avoided.

Using Statistical Bias for Prediction

The statistical biases described in the preceding sections may be used for purposes of prediction in answering questions of biological importance, such as the following: 'Does this DNA region code for a protein?' 'Which is the coding strand in this DNA sequence?' 'Which is the coding phase?' 'Does this sequence contain an error?' 'What are the boundaries between introns and exons?' 'Is this gene strongly expressed?'

In previous study, sequence characteristics were investigated according to *signals* determined by specific patterns. In this section, they will be pursued by identifying their *contents*; *i.e.*, by determining the statistical properties of their distribution as a function of the bases that constitute them.

Search for coding sequences

The most powerful method for determining whether a stretch of DNA is a coding region, and which relies on the fewest *a priori* hypotheses, consists in seeking period 3 irregularities in nucleotide distribution. Non-uniform codon usage within a gene or exon should be revealed by period 3 bias in the frequencies of occurrence of individual nucleotides. This method requires no prior knowledge of the codon usage table for the organism, or even of the genetic code. The frequency of occurrence of each base at positions 3i, 3i + 1, and 3i + 2 are simply calculated and compared with the average frequency of occurrence in the sequence.

$$\Delta = \sum_{N=A,T,G,C} \sum_{3\,phases} \left| f_{N/phase\,i} - f_N \right|$$

The value of Δ is calculated within a window of around 100 nucleotides moved progressively along the sequence being studied. If the value of Δ is traced as a function of the position of the window, a profile is obtained whose peaks fairly well represent the coding regions.

This method is generally well adapted to seeking introns and exons in eukaryotic genomic sequences. Splice site consensus sequences are usually insufficient to accurately determine intron/exon boundaries, for which the content research method can be an effective complement.

Coding phase analysis

The above technique may be considerably refined by utilizing the codon usage table for the species from which a sequence derives, thereby allowing prediction of which of the three phases is coding. The principle is to compare the frequencies of triplets appearing in the three phases with the canonical triplet frequencies, then to determine which of the former are most similar to the latter.

Practically speaking, the frequency of occurrence f_1 of N codons in a window on phase 1 is compared with the individual frequencies of codons in the standard genetic code usage table:

$$f_1 = \prod_{i=1}^{N} f_{codon\ i}$$

By shifting first one and then two nucleotides, the calculation is repeated for phases 2 and 3, which gives two other frequencies, f_2 and f_3. The probability of each phase being the coding phase is given by Bayes' formula:

$$p_1 = f_1/(f_1 + f_2 + f_3)$$
$$p_2 = f_2/(f_1 + f_2 + f_3)$$
$$p_3 = f_3/(f_1 + f_2 + f_3)$$

By displacing the length N window along the sequence, it is possible to trace the profiles of probabilities p_1, p_2, and p_3, whose peaks indicate the positions of genes that code for proteins with remarkable precision. This method is very sensitive and can be used with shorter windows (around 12 codons). It is extremely useful for detecting insertion and deletion sequencing errors, which can result in a reading frame shift and thus easily be detected by using the probability graph, p_1, p_2, and p_3.

Gene expression level

The expression level of a gene may be estimated by comparing its codon usage frequency with the standard codon frequencies given in the codon usage table for the species. The gene expression level may be quantified using a quantity known as the *codon adaptation index* (CAI), which is calculated. Each codon i contained in the gene is assigned a score w_i equal to the ratio of its frequency to the frequency of the most frequent codon that codes for the same amino acid. If codon i is the most frequent, then w_i equals 1. For a codon that is systematically avoided, w_i is close to 0. The codon adaptation index is the geometric mean of the w_i scores of the set L of the gene's codons.

$$Index = \left(\prod_{i=1}^{L} w_i \right)^{\frac{1}{L}}$$

The CAI score obtained ranges between 0 and 1, increasing as the gene conforms to the standard utilization frequency of the species' genetic code. For example, protein genes that are strongly expressed in the yeast, such as ribosomal proteins and histones, have scores of between 0.52 and 0.92, whereas regulatory protein genes, of which there are only a few copies per cell, have a score of 0.1.

The CAI may be used to estimate the expression level of a gene whose function is unknown. It is also useful when expressing a recombinant protein in a heterologous host, for example, a human protein in a bacterium. The CAI for a human gene, in combination with the bacterial codon usage table, allows us to predict whether a gene will be efficiently expressed, and may be used to guide the modification of certain codons in order to better adapt the gene to its new host.

Modeling DNA Sequences

In his novel, 'Jurassic Park', the American author, Michael Crichton, tells the story of a molecular biologist, a certain Dr. Wu, who sequences the DNA of dinosaurs preserved in yellow amber in order to resuscitate the extinct animals. To support his fiction, the author even went to the extent of devising a figure that supposedly represented a ~ 1.5 kbase fragment of dinosaur genome!

Is this dinosaur DNA sequencing credible? For the connoisseur of biomathematics, it does not withstand scientific scrutiny, hence it is just pseudoscience. Subjecting the fictional sequence to the statistical tests described above reveals that it does not present the biases expected for vertebrate DNA, such as the 40% G + C level and low CG dinucleotide frequency But what *would* a novelist or biological faker have to do to concoct a pseudosequence capable of mystifying the shrewdest specialist?

Clearly, the solution would be to find a method that generated sequences displaying exactly the same statistical biases as real biological ones.

Markov chains

A very simple first approach is to attempt to reproduce both the nucleotide composition of the genome and its CG nucleotide content.

Elementary statistical analysis of a body of vertebrate genomic sequences yields the following results for the relative frequencies of nucleotides:

$$f_A = 0.30, f_T = 0.29, f_C = 0.21, f_G = 0.20$$

In nucleotide frequencies, the frequency of the CG dinucleotide (in boldface) is observed to be lower than that of the others:

A very simple approach for generating a sequence that reproduces these frequencies involves iteration, by adding one nucleotide at a time. Suppose that a C has just been added at position i. Using the above table, it is easy to calculate the probabilities of encountering A, C, G, or T after that C. For example, the probability for an A to follow a C is:

$$p(A/C) = \frac{f_{CA}}{f_c} = \frac{f_{CA}}{f_{CA} + f_{CC} + f_{CG} + f_{CT}}$$

Equivalent probabilities may be symmetrically calculated for the other three nucleotides, C, G, and T. They are indicated in **Figure** for the example of vertebrate sequences, expressed in percentages. Within rounding-off errors, line totals should equal 1 (100 percent).

Utilizing these probabilities in conjunction with a random number generator, it is possible to progressively generate a pseudosequence that mimics the biases observed in vertebrate sequences. This method of random fabrication using a probability Utilizing these probabilities in conjunction with a random number generator, it is possible table to generate state $i + 1$ from state i is called a *Markov chain* (or Markov *process*), well known in the field of applied mathematics. If state i is called e_i, the probability $p(e_{i+1}|e_i)$ that i will go to state e_{i+1} is given by the Markov process transition table, as indicated above. A simple Markov chain is a discrete random process that takes only the immediately preceding state into account in generating the current state.

Higher-order biases

The simple Markov chain just described is a perfect tool for reproducing DNA composition in terms of mono- and dinucleotides. Nevertheless, as seen above, higher-order irregularities exist in genomes, in particular, period 3 biases related to the translation of the genetic message by codons (*i.e.*,. nucleotide triplets). To simulate these biases and produce 'plausible' sequences, we can use an extension of simple Markov chains. All that is required to produce state i is to utilize more complex probability tables that not only take state $i - 1$ into account, but also states $i - 2$, $i - 3 \ldots i - k$. In this case, we speak of an 'order k Markov process'. State e_i is then produced with probability $p(e_i|e_{i-1}...e_{i-k})$. With a Markov process of order $n - 1$, it is possible to produce pseudosequences that reproduce the frequencies of the k-tuples of nucleotides of a family of biological sequences for all k between 1 and n.

Using Markov processes to study DNA sequences

Up to this point, the objective has been to produce random pseudosequences that would fool a biologist, which might seem to be of quite trivial interest. Indeed, the utility of Markov chains lies elsewhere, since they are powerful probprobabilistic tools for testing hypotheses concerning the nature of DNA sequences. Using a Markov transition table makes possible *a posteriori* computation of the probability that a given sequence $e_1, e_2 ... e_n$ could have been produced by the corresponding Markov process. This equals the product of the probabilities of the individual transitions:

$$prob(e_1 e_2 \ldots e_n) = \prod_{1>k\leq n} p(e_k | e_{k-1})$$

For example, in the above model, it is possible to calculate the probability of sequence CGCG and to compare it with probability of AATG. The first sequence, which is rich in CG, is thus shown to be much less probable than the second:

prob (CGCG) $= 0.05 \times 0.22 \times 0.05 = 5.5 \times 10^{-4}$

prob (AATG) $= 0.34 \times 0.25 \times 0.26 = 2.2 \times 10^{-2}$

This kind of calculation is useful when comparing various hypotheses for the sequence of a given DNA segment, as long as it is possible to associate a Markov process with each hypothesis envisaged. The following is a concrete example illustrating this type of application.

A biologist has just determined some cDNA sequences obtained from messenger RNA derived from a mammalian cell culture. Even when all the required precautions are taken, cell cultures are sometimes contaminated by intracellular bacterial parasites known as mycoplasms. The RNA extraction process does not permit separating the nucleic acids of the mammalian cells from those of the mycoplasms. At the end of the experiment, some of the sequences obtained therefore could have come from the mycoplasms rather than from the mammalian cells. How can the intrusive sequences be distinguished from the ones of interest?

Several mycoplasm genomes have already been completely sequenced, therefore it is possible to study their dinucleotide composition and to deduce a Markov process transition probability table for them, as was done above for the vertebrate sequences.

Given a fragment of a sequence obtained by our biologist, it is possible to calculate the probability that it was produced by Markov processes, utilizing the probabilities derived from either the vertebrate or the mycoplasm genomes. Comparing these two probabilities, it is often also possible to decide which hypothesis is the most likely. Consider the following sequence as an example:

S = 5′ - TTTCAAATAATCGTGAAATATCTTC–3′

A posteriori calculation utilizing the two transition tables (vertebrate and mycoplasm) given above yields the following results:

$$p_{\text{vertebrate}}(S) = 4.3 \times 10^{-15}$$
$$p_{\text{mycoplasm}}(S) = 18.7 \times 10^{-15}$$

Despite the short length of this sequence (only 25 nucleotides), there is a rather clear difference, since the Markov process based on mycoplasm DNA produces a result that is four times more probable than the Markov process based on vertebrate DNA. This sequence may therefore be attributed to contamination with a mycoplasm bacterium. The differences are even clearer (by several orders of magnitude) in longer sequences, usually eliminating any ambiguity.

Complex Models

When examining protein-coding regions in a given organism, it is preferable to utilize the codon usage table for the species, for the human and for *E. coli*. However, the simple Markov processes described above do not allow this to be done directly, since codon usage tables only consider triplets that are in phase with reading frames that code for proteins. Triplets that are shifted +1 or +2 nucleotides with respect to the reading frame are not taken into account, unlike in a simple Markov process. For sequences that encode proteins, it is possible to compile triplet frequency tables for both the correct phase (phase 0), which corresponds to the classical usage table of the genetic code, and for the two other phases (phase +1 and phase +2).

These three tables are rather complicated, but carefully examining them reveals certain interesting features: While some triplets, such as TAG and GGG, are systematically lacking in all three phases, others have very different frequencies, according to the phase considered. For example, whereas GAA is relatively abundant in the coding phase (4.05%), it is rare in phase +1 (only 0.36%). This observation calls for two important remarks:

1. The simple Markov processes discussed above cannot capture the full complexity of biases imposed by using the genetic code within coding regions.
2. To predict the coding phase, it must be possible to use biases that are not the same among the three phases.

It is nevertheless possible to devise a sequence-coding model that includes the notion of phase. This requires construction of a kind of Markov process that utilizes three probability tables deduced from the frequencies in rotation. During each cycle, this process generates a nucleotide that shifts its reading phase one position.

This is a modified order 2 Markov process, because knowledge of the two preceding nucleotides, e_{i-1} and e_{i-2}, is used to complete the nucleotide triplet by referring to probability tables: $p_0(e_i|e_{i-1},e_{i-2})$, $p_{+1}(e_i|e_{i-1},e_{i-2})$ and $p_{+2}(e_i|e_{i-1},e_{i-2})$ for each of the three phases, 0, +1, and +2. This Markov process could be considered an automaton, permitting fabrication of very convincing *E. coli* coding pseudosequences. These pseudosequences also reproduce the preferential codon usage of this bacterium by utilizing the phase 0 probability table, as well as the context effects described above, which are captured by +1 phase probabilities for the 3′ context and +2 phase probabilities for the 52 context. An example of a generated sequence is:

ATGCTATTCAGCTTCATCCTGACAAAACCGGGGGACGGTA ACGAGGCGCTGATCTATATC...

MetLeuPheSerPheIleLeuThrLysProGlyAspGlyAsnGluAlaLeuIle TyrIle...

As for classical Markov processes, the main utility of this type of model obviously is not to produce pseudosequences, but rather to test *a posteriori* whether a biological sequence does indeed correspond to a coding phase. To do this, it is possible, as shown above, to calculate the probability that a given sequence had been produced by the modified Markov process:

$$prob(e_1e_2\ldots e_n) = \prod_{2<k\leq n} p_{k \bmod 3}(e_k|e_{k-1},e_{k-2})$$

This probability evaluation alone does not provide much information, but it is interesting to compare it with other probabilities. As discussed above, a direct application is prediction of the species from which a sequence derives; for example, determining whether a sequence has been contaminated by exogenous genetic material. To do this requires calculating and comparing the probabilities, using Markov processes adapted to various frequencies in the organisms studied.

Another interesting application concerns searching for the coding phase. Given a cDNA sequence presumed to code for a protein, is it possible to determine which is the coding phase? The following three probabilities may be calculated, each of which is shifted by one nucleotide with respect to the preceding one, and which correspond to the three phases, 0, +1, and +2:

$$prob(e_1 e_2 \dots e_n); prob(e_2 e_3 \dots e_{n+1}); prob(e_3 e_4 \dots e_{n+2})$$

Their comparison usually allows unambiguous prediction of which of the three phases is coding. In the following example, the 25 first codons of a natural *E. coli* gene are:

ATGAAAGGCGGAAAACGAGTTCAAACGGCGCGCCCTAAC
CGTATCAATGGCGAAATTCGCGCCCCAGGAAGTTCG

In spite of the short length of this sequence, the results obtained using the modified Markov model described above are very clear: the correct phase (phase 0) has a probability between ten and one-hundred million times greater than the other two phases:

$$prob(\text{phase } 0) \approx 10^{-97}; prob(\text{phase} + 1) \approx 10^{-105}; prob(\text{phase} + 2) \approx 10^{-104}$$

This result is all the more remarkable in that it uses only statistical composition information and does not directly depend on the presence of start or stop codons, which, as seen in the preceding chapter, is necessary for pattern recognition. This approach is therefore very robust and may be applied within a fragment of any coding sequence. It can also be applied to very short open reading frames (a few dozen codons) to predict whether they are in fact real genes coding for small proteins or peptides.

Sequencing Errors and Hidden Markov Models

Recognition of coding phases is a fundamental problem faced by biologists attempting to annotate genomes. Despite the technical progress that has been made in sequencing methods, errors inevitably slip into raw data. When this concerns the insertion or deletion of a nucleotide, the result is a shift in the DNA reading frame. The pattern recognition methods mentioned in the preceding chapter cannot be used to solve this type of problem, which results in possible failure to detect a gene in a genome.

The sensitivity of the statistical approach provided by Markov processes in locating coding phases makes it possible to detect this type of error. In explaining how to proceed, the following section will again hypothesize using a Markov-type process to fabricate false genomic sequences containing errors.

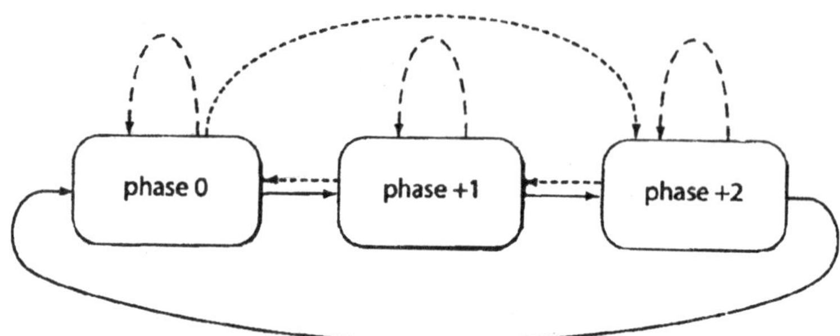

Fig. 4.7. Schematic representation of the modified model, allowing phase jumps.

Sequencing errors are relatively rare, with an incidence of around 1 insertion or deletion per 1,000 nucleotides in the raw data. The model may be used, in which the phases are incrementally cycled, utilizing the associated probability table each time. The only modification that will be introduced is that the system will occasionally be allowed to 'slip'; *i.e.*, jump to a phase other than the one expected. The new diagram of the process presents frequent transitions (solid lines), which correspond to those of the correct model, as well as rare transitions, (dotted lines), which correspond to insertions/deletions.

This system is a probabilistic automaton consisting of 48 possible states. In fact, each transition depends on the two preceding nucleotides e_{-1} and e_{i-2}, which represent 16 different dinucleotides. These 16 possibilities combined with the three phases amount to 48 states. The automaton goes from one state to another, as a function of its probability tables. At each transition, it also produces a nucleotide that is added to the end of the sequence being fabricated.

The objective now is to utilize this model for *a posteriori* calculation of the probability of a given sequence, as was done earlier for classical Markovian processes. However, this cannot be done, since if only the final sequence is available, it is impossible to reconstitute the route. In fact, unlike all the automata constructed earlier in the chapter, in this one, several different transitions can yield the same nucleotide at each step, according to whether a normal transition or an insertion/deletion occurs. For this automaton, it all depends on whether the current state is in phase 0, +1, or +2. Such a model, in which the sequence produced does not allow reconstruction of the progression of events of the automaton, is called a *hidden Markov process*, precisely because its states are unknown.

For hidden Markov processes, there is a very large number of possible routes, therefore a series of possible states yielding the same sequence. In the above example, given the three phases, 3^{n-2} routes exist for a sequence of length n. However, the probabilities associated with them may be very different. In particular, since transitions associated with phase jumps (insertions/deletions) have very low probabilities, the routes that take them are usually highly unlikely.

The problem at hand is to identify the coding phase and any shifts in it caused by sequencing errors. This information corresponds to the sequence of the states of the automaton. According to which criteria should the route be selected among the 3^{n-2} possible? The overall probability of each route may be calculated by computing the individual probabilities associated with each transition as it proceeds. The route with the highest probability can then be selected in order to reconstruct the one with the most plausible sequence of states.

Viterbi algorithm

There are obviously too many probabilities associated with each route to calculate them exhaustively. The Viterbi algorithm avoids this difficulty by using a dynamic programming approach. For a sequence $e_1e_2\ldots e_n$ of length n, it begins by calculating the probability of the best route from e_1 to e_k for k progressively varying between 1 and n and terminating in each of the possible states of the automaton. In the example of the search for coding phase, this amounts to seeking the probability of the best route over the k first mucleotides and terminating in either phase 0, phase +1, or phase +2. A table P of dimension $3 \times (n-1)$ containing all these possibilities is drafted. The table is initialized at 1 in the three phases for the second nucleotide e_2. By recurrence, it is then possible to calculate the kth column of table P:

$$P(i,k) = \max_{i=0,1,2}(P(j,k-1)p_{j,i}(e_k|e_{k-1},e_{k-2}))$$

Variable i and j represent the index of the phase and $p_{j,i}(e_i|e_{k-1}, e_{k-2})$ represents the probability of transition of the dinucleotide $e_{k-2}e_{k-1}$

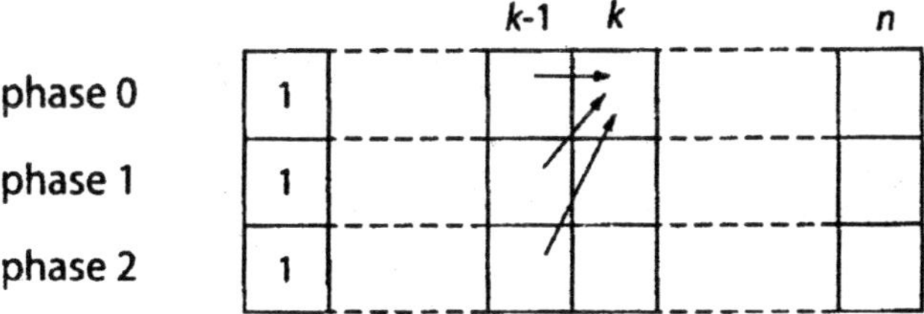

Fig. 4.8. Completion of table P. for each cell, the probabilities of the three possible transitions are evaluated, then the maximal one is selected.

toward the nucleotide e_k passing through phase j toward phase i. If $i = j + 1$ modulo 3, then it is a standard transition, corresponding to the solid-stem arrow, and the probability can be calculated from the tables. In the contrary case, the transition corresponds to an insertion or deletion and the probability is lower, on the order of the error level, estimated to be around 10^{-3}.

When the table is complete, the last (n^{th}) column indicates which of the three phases has the highest probability. The inverse route is then taken in the table in order to determine the path it followed. For each column k, this amounts to finding the value of j in the above equation corresponding to the maximum of the three terms. This value of j indicates the most probable state of the hidden Markov process at the level of the $k - 1^{th}$ nucleotide. The list of the most probable states as a function of k yields a prediction of the correct coding phase.

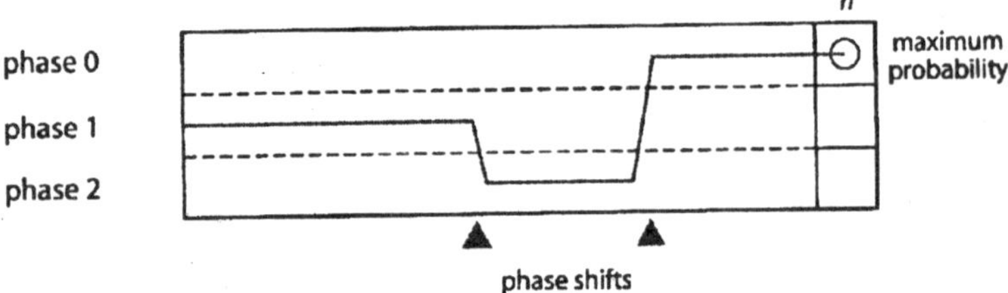

Fig. 4.9. Reverse reconstitution of the path across table P, starting from the final maximum probability score.

In practice, it is not these probabilities that are calculated directly, but their logarithms, which has two advantages: (i) calculation of a *product* is replaced by calculation of a *sum*, which is faster, since all that is necessary is to add the logs of the transition probabilities $\log(p_{j,i}(e_i \mid e_{k-1}, e_{k-2}))$; (ii) using logarithms avoids the risk of exceeding the capacity of the machine (underflow), since the overall probability of long sequences rapidly becomes ridiculously small.

Predictions carried out using the Viterbi algorithm are often very accurate and of much higher quality than those produced using profile score methods. In the following example, which uses the nucleotide sequence of an *E. coli* gene, two errors have been introduced intentionally: one deletion and one insertion. This kind of double error can be especially difficult to detect, since the effects on the phase mutually compensate for each other such that the length of the reading frame is not affected. Nevertheless, the coding sequence between the two errors is profoundly affected. However, the Viterbi algorithm nearly perfectly detects the two ill-timed changes in the sequence.

ATTAAAGGCGGAAAACGAGTTCAAACGGCGCGCCCTAACCGTATCAATGG

CGAAATTCGCGCCCCAGGAAGTTCGCTTAACAGGTCTGGAAGGCGAGCAG

CTTGGTATTGTGAGTCTGAGAGAAGCTCTGGAGAAAGCAGAAGAAGCCGG

AGTAGACTTAGTCGAGATCAGCCTAACGCCGAGCCGCCGGTTTGTCGTAT

AATGGATTACGGCAAATTCCTCTATGAAAAGAGCAAGTCTTCTAAGGAAC

Fig. 4.10. Sequence of a fragment of the Escherichia coli infC gene with two phase shifts; first a deletion, then an insertion.

Hidden Markov Processes: a General Sequence Analysis Tool

The phase shift search discussed above is just one relatively simple example developed to illustrate the material treated in this chapter. However, the range of Markov model applications in the domain of biological sequence analysis is much greater. Refining this type of model and having it take into account frequencies in non-coding regions and in start and stop codons can convert it into an automatic genome gene-seeking tool. In eukaryote genomes, in which coding regions are interrupted by introns, statistical analyses have been conducted in both intron and exon regions and the frequencies obtained used to construct hidden Markov processes. The most probable route reconstituted using the Viterbi algorithm may then be utilized to predict spliced regions in messenger RNA. Finally, hidden Markov processes have also been used with protein sequences to predict secondary structures (α-coils and β-sheets).

Search for Genes – a Difficult Art

The exhaustive search for all the genes contained in a large genome is a complex task. In higher eukaryotes, it is complicated by the 'dilution' of relevant information in non-coding sequences of repetitive DNA and in intergene regions. The presence of lengthy introns separating short exons renders precise assembly of coding zones rather delicate. In such case, none of the methods discussed until now is by itself sufficient to identify genes. In general, several different approaches are necessary to achieve a reliable result. Currently used approaches are based on the following strategy:

1. Predict exon and intron regions, using a statistical method, most often a hidden Markov process.
2. Search for consensus patterns of donor and acceptor splicing sites, using a finite automaton method.
3. Combine the two types of information in order to precisely predict intron and exon borders.

4. Assemble the predicted exons and compare the sequence obtained with cDNA and EST sequence databases of the same organism. All or part of the predicted sequence may be found in one of these sequences, which derive from messenger RNA.
5. If this search fails, it is still possible to compare the protein sequences predicted from the genomic DNA (translated into all possible phases) with protein sequence databases. If the genomic DNA region being considered effectively codes for a protein that possesses a homolog identified in another species, a BLAST-type search based on translated sequence fragments may be carried out to identify these homologies. This would permit identifying or confirming some coding parts, thus some intronic parts of the gene.

Several very elaborate tools that combine several of these approaches are accessible on the web, for example, GenScan, which has been utilized to analyze and document the entire human genome.

5

DATABASES

The origin of bioinformatics was associated with the need for maintenance of databases containing biological, biochemical, and clinical data. Historically, the process of collecting bioinformatic sequences started with amino acid sequences in proteins and then proceeded by coordinating and standardizing submissions by the institutions engaged in maintaining the database. Currently there is fast growth in the volume of the data stored in bioinformatic databases. Bioinformatic databases are also growing in number. There are dense links between different databases, since they often contain information pertaining to several aspects of, for example, the same sequence of amino acids in a protein. Also, numerous Internet resources offer services related to searching and browsing bioinformatic databases and to various algorithms for data processing and inference based on bioinformatic data.

In this book, we referenced many databases containing bioinformatic data and we have shown samples of their resources. In this short chapter we present a view of the types and structures of bioinformatic databases and other bioinformatic resources and on the relations between them. We provide a list of bioinformatic Web sites and a classification of the types of bioinformatic databases and bioinformatic resources, based on their content and functions. Our list of databases and bioinformatic sites is not comprehensive; rather, it contains those most widely known and some samples of others. Also, our view on their classification may not be unquestionable, owing to the large variety of data and functions and the dense links between databases. Our aim when presenting the overview below was to provide a picture of the Internet resources related to bioinformatics.

In the references, we have provided Internet addresses of many databases and other bioinformatic sites. However, they should be treated with caution owing to the to constant evolution in the field.

Genomic Databases

The best known genomic database is GenBank, the database of genomic sequences maintained by the NCBI. It contains all annotated nucleic acid and amino acid sequences. Its contents are mirrored by two other databases, the EMBL database and DDBJ. Apart from presenting and annotating sequences, these databases offer many functions related to searching and browsing sequences. They perform services concerning submitting new sequences to GenBank, and also contain links to various bioinformatic internet sites.

Many databases contain more specialized information on genomic sequences. Examples of such databases are the Single Nucleotide Polymorphisms (SNP) Consortium database for biomedical research, the databases of highly conserved DNA motifs, and the cisRED database of gene promoter and regulatory sequences. The database serves for standardizing the nomenclature for genes.

Proteomic Databases

Owing to the correspondences between amino acid and codon sequences, there are strong links between protein and nucleotide databases. As mentioned above, amino acid sequences of proteins are available at GenBank along with nucleotide sequences. Data on sequences of amino acids, on the taxonomy, functional aspects of proteins, protein families and domains, as well as data on known secondary and 3D structures of proteins, are stored in proteomic databases, databases, Swiss-Prot, Uni-Prot, and ExPASy (*Expert Protein Analysis System*) comprise information including the function, classification, amino acid sequences in proteins and structures of proteins. The PDB (*Protein Data Bank*) database contains annotated data on the spatial structures of proteins and biological macromolecules. It also includes data on their sequences, functions, and related diseases.

There are also many databases specializing in particular aspects of proteins, protein functions, experiments involving proteins etc., such as databases of protein 2D gels, restriction enzymes, and secondary structures.

RNA Databases

Information on sequences of ribonucleotides in RNA, coding and noncoding RNA sequences, the functions of RNA molecules and their spatial structures, is available in the databases. Rfam database stores noncoding RNA (ncRNA) families. Rfam also contains multiple sequence alignments and covariance models. The GtRNA database stores genomic tRNA ribonucleotide sequences and secondary structures. The Jena index of RNA structures provides a lot of information about RNA, including indexes of the locations of molecular structures in the PDB database. Data on ribonucleic acid sequences in RNA can also be found in GenBank.

Gene Expression Databases

Numerous databases contain data on expression levels measured in various experiments. They are aimed at making possible the sharing of data in the new field of microarray experiments. A part of the NCBI database, NCBI Gene Expression Omnibus (GEO), is a database including links to microarray-based experiments measuring mRNA, genomic DNA, and protein abundances, as well as non array techniques such as serial analysis of gene expression (SAGE), and mass spectrometric proteomic data. The MGED database contains datasets from many experimental studies involving gene expression. It also contains links to gene expression data-processing procedures and ontologies. The databases CGED (Cancer Gene Expression Database) and ONCOMINE deliver published cancer gene expression data to the research community.

Ontology Databases

Probably the most extensively used ontology database is GO (*Gene Ontology*), which provides controlled vocabularies for supporting analyses of gene expression measurements and other molecular-biology experiments. However, other ontologies are also developing at a fast rate. One database containing links to many Internet ontologies is OBO (*Open Biomedical Ontologies*). It provides Web addresses of many sites containing biomedical structured vocabularies, including GO, the Generic Model Organism Project (GMOD), Microarray Gene Expression Data (MGED), The National Center for Biomedical Ontology vocabulary.

Databases of Genetic and Proteomic Pathways

The area of genetic pathways is a very fast-growing field for bioinformatic data. One large database containing data on the genetic pathways is KEGG, the Kyoto Encyclopedia of Genes and Genomes, a Web site that organizes databases and associated software. The BioCarta database supports proteomic studies by providing information on proteomic pathways, as well as on reagents, antibodies, proteins, cells and cell-based assays. Pathway databases offer a graphical presentation of their contents, which provides a useful support for qualitative understanding of signaling, regulatory, and other mechanisms.

Programs and Services

A large variety of programs for performing bioinformatic computations are available on the Internet. Here we list some areas and related Web sites.

The Internet bioinformatic services that are most important to biologists and most frequently used by them are probably those for aligning sequences against sequence databases. This task is performed by variants of the BLAST program, or earlier program FASTA. Sequence databases contain links to these programs. Related to this area are also programs for performing multiple (block) alignments of sequences, e.g., CLUSTAL W.

A well-known program and Internet site for inferring phylogenetic trees is PHYLIP. The Web site also includes addresses of other Internet resources related to phylogenetics.

Examples of programs for annotating genome sequences are GENESCAN or MEME. Also, the NCBI Web site contains a simple service for searching of open reading frames.

Several servers are aimed at performing assembly of DNA sequences from reads, for example, Atlas Genome Assembly, Arachne, Celera Assembler, Jazz, Phusion, PCAP, and Euler.

There are many programs concerned with molecular geometry and visualization, for example, 3DNA for nucleic acid structures or more general program for showing biomolecules Ras Mol.

There are many different programs related to algorithms related to gene expression data, including image processing, normalization, classification and clustering, and searches of ontologies.

Clinical Databases

In this last section we mention repositories of clinical data. These databases are growing at a very fast rate owing to their importance in developing knowledge about disease etiologies, therapies, treatments, and so forth. They collect data describing clinical cases, diagnoses, therapy protocols, recovery, and survival. Different types of data are brought together: patients records, diagnostic tests including medical images, treatment plans and follow up data on patient cohorts concerning recovery, complications and survival. There are many aspects of the development of clinical databases: quality, completeness and volume of data, availability and so forth. An example of a clinical database is the European EURODIAB database related to type 1, childhood diabetes, resulting from many years of collecting clinical cases of type 1 diabetes mellitus. Along with storing patient records, treatments, and survival data, clinical databases can also organize and help with access to tissue banks. An example is the GENEPI (GENEtic Pathways for the Prediction of the Effects of Ionising Radiation) database, which stores clinical data concerning therapeutic radiation and, at the same time, can help in arranging access to tissue samples.

Protein Data Bank

The *Protein Data Bank* (PDB) was established at Brookhaven National Laboratory (BNL) in 1971 as an archive for biological macromolecular crystal structures. Nobel prizes have been awarded for the determination and analysis of some of the structures in the PDB. It represents one of the earliest community-driven molecular biology data collections. In the beginning the archive held seven structures, and with each passing year a handful more were deposited. In the 1980s, the number of deposited structures began to increase dramatically. This increase was due to the improvements in technology for all aspects of the crystallographic process, the addition of structures determined by *nuclear magnetic resonance* (NMR) methods, and changes in community views about data sharing. By the early 1990s, the majority of journals required a PDBidentification (PDBid) for publication and at least one funding agency, the National Institute of General Medical Sciences (NIGMS), adopted the guidelines published by the International Union of Crystallography requiring data deposition for all structures determined using NIGMS funds. At the beginning of 2002 there were more than 17,000 entries in the archive.

Accompanying this rapid growth, the mode of access to PDB data has changed over the years as a result of improved technology. Data distribution is now primarily via the World Wide Web (www) rather than via magnetic media. Further, the need to analyze diverse subsets of the data has led to the development of modern data management systems.

Initial use of the PDB had been limited to a small group of experts involved in structural biology research. Today depositors to the PDB have expertise in the techniques of X-ray crystal structure determination, NMR, cryoelectron microscopy, and theoretical modeling. PDB users are a very diverse group of researchers in biology and chemistry, as well as educators and students at all levels. The tremendous influx of data soon to be fueled by the structural genomics initiative, and the increased recognition of the value of these data toward understanding biological function, continually demand new ways to collect, organize, and distribute the data.

Since October 1998, the PDB has been managed by the Research Collaboratory for Structural Bioinformatics (RCSB), which is a consortium consisting of Rutgers, the State University of New Jersey; the San Diego Supercomputer Center at the University of California, San Diego; and the National Institute of Standards and Technology. In this chapter, we describe the current procedures for collecting, validating, annotating and distributing PDB data. Finally, we describe the plans for further automating and improving the PDB so it can meet emerging challenges posed by researchers and educators in the field of structural bioinformatics.

Data Acquisition and Processing

A key component of the PDB is the efficient capture and curation of the data—data processing. Data processing consists of data deposition, annotation, and validation.

In the present system, data (atomic coordinates, structure factors, and NMR restraints) may be submitted via e-mail or via the Web-based AutoDep Input Tool (ADIT) developed by the RCSB PDB. ADIT is built on top of the macromolecular Crystallographic Information File (mmCIF) dictionary that contains 1700 terms that define the macromolecular structure and the crystallographic experiment. The mmCIF dictionary has been further extended to form the PDB exchange dictionary, which includes terms needed for tracking and other information management purposes. ADIT is complemented by MAXIT (MAcromolecular EXchange Input Tool), a program that performs many of the data-processing tasks and checks. This integrated system helps to ensure that the data submitted are consistent with the mmCIF dictionary, which defines data types, enumerates ranges of allowable values where possible, and describes allowable relationships between data values.

Each deposition to the PDB is represented by the PDBid—a four character code of the form nXYZ, where n is an integer and X, Y, and Z are alphanumeric characters, for example, 4HHB. The PDBid is assigned arbitrarily and is an immutable reference to the structure, and indeed is the only absolute way of retrieving a desired structure from the PDB, although this shortcoming is being addressed. PDBids are never reused and remain the link between the structure and the literature reference that describes that structure.

After a structure has been deposited using ADIT, a PDBid is automatically and immediately sent to the author. This procedure is the first stage in which information about the structure is loaded into the internal core database, validated, and annotated. This step involves using ADIT to help diagnose errors or inconsistencies in the files. The completely annotated entry as it will appear in the PDB resource, together with the validation information, is sent back to the depositor. After reviewing the processed file, the author sends any revisions. Depending on the nature of these revisions, steps 2 and 3 may be repeated. Once approval is received from the author, the entry and the tables in the internal core database are ready for distribution. The schema of this core database is a subset of the conceptual schema specified by the mmCIF dictionary.

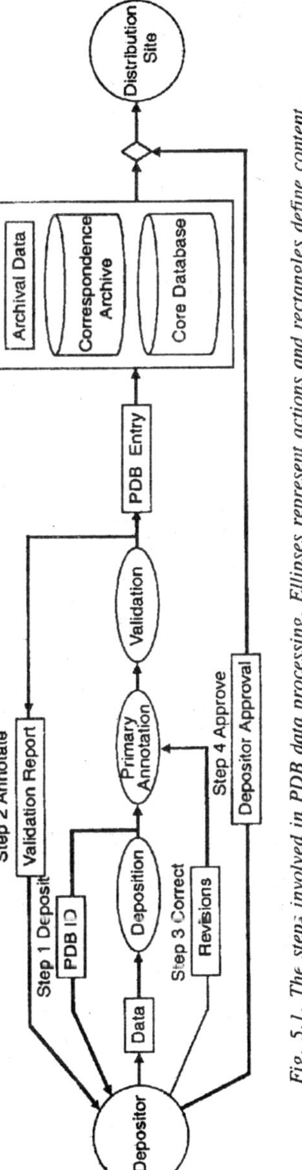

Fig. 5.1. The steps involved in PDB data processing. Ellipses represent actions and rectangles define content.

All aspects of data processing, including communications with the author, are recorded and stored in the electronic correspondence archive. This record makes it possible for the PDB staff to retrieve information about any deposited entry. Current status information, including the entry's authors, title, and release status, is stored for each entry in the core database and is made accessible for query via the WWW interface. Entries before release are categorized as "in processing" (PROC), "in depositor review" (WAIT), "to be held until publication" (HPUB) or "on hold until a depositor specified date" (HOLD).

Content of the Data Collected by the PDB

All the data collected from depositors by the PDB are considered primary data. Primary data contain, in addition to the coordinates, general information required for all deposited structures and information specific to the method of structure determination. Table 5.1 contains the general information that the PDB collects for all structures as well as the information specific to X-ray and NMR experiments.

Table 5.1. Content of Data in the PDB

Content of All Depositions (X-ray and NMR)

- *Source*: Specifications such as genus, species, strain, or variant of gene (cloned or synthetic); expression vector and host, or description of method of chemical synthesis
- *Sequence*: Full sequence of all macromolecular components
- Chemical structure of cofactors and prosthetic groups
- Names of all components of the structure
- Qualitative description of the characteristics of the structure
- Literature citations for the structure submitted
- Three-dimensional coordinates

Additional Items for X-ray Structure Determinations

- Temperature factors and occupancies assigned to each atom
- Crystallization conditions, including pH, temperature, solvents, salts, methods Crystal data, including the unit cell dimensions and space group
- Presence of noncrystallographic symmetry
- Data collection information describing the methods used to collect the diffraction data including instrument, wavelength, temperature, and processing programs
- Data collection statistics including data coverage, $R_{sym.}$ data above 1, 2, 3 sigma levels and resolution limits
- Refinement information including R factor, resolution limits, number of reflections, method of refinement, sigma cutoff, geometry rmsd, sigma
- Structure factors: h, k, l, Fobs, σ Fobs

Additional Items for NMR Structure Determinations

- For an ensemble, the model number for each coordinate set that is deposited and an indication if one should be designated as a representative
- Data collection information describing the types of methods used, instrumentation, magnetic field strength, console, probe head, sample tube

Sample conditions, including solvent, macromolecule concentration ranges, concentration ranges of buffers, salts, antibacterial agents, other components, isotopic composition

Experimental conditions, including temperature, pH, pressure, and oxidation state of structure determination and estimates of uncertainties in these values

Noncovalent heterogeneity of sample, including self-aggregation, partial isotope exchange, conformational heterogeneity resulting in slow chemical exchange

Chemical heterogeneity of the sample (e.g., evidence for deamidation or minor covalent species)

A list of NMR experiments used to determine the structure including those used to determine resonance assignments, NOE/ROE data, dynamical data, scalar coupling constants, and those used to infer hydrogen bonds and bound ligands. The relationship of these experiments to the constraint files are given explicitly

Constraint files used to derive the structure as described in Task Force recommendations

---

Historically, NMR data have been placed in a format defined around crystallographic information. The PDB is currently working with an NMR Task Force and the BioMagResBank (BMRB) to develop an NMR data dictionary as an extension to the mmCIF. This dictionary includes descriptions of the solution components, the experimental conditions, enumerated lists of the instruments used, and information about structure refinement. This dictionary will be used as deposition and validation tools specifically for NMR structures. NMR coordinate data are currently deposited with the PDB and other NMR-specific experimental data are deposited with the BMRB. Plans are in place to have a single interface for the deposition of data to both the BMRB and the PDB.

The information content of data submitted by the depositor is likely to change as new methods for data collection, structure determination, and refinement evolve. A case in point is the need for structural genomics projects to collect all information that would be in the material and methods section of a paper describing a structure. The ways in which these data are captured are also changing as the software for structure determination and refinement evolve to produce the necessary data items as part of their output. The ontology-driven approach to software development used by the PDB makes it simple to collect new items of data once they are described in the mmCIF or extension dictionary.

Validation and Annotation

Validation refers to the procedure for assessing the quality of deposited atomic models (structure validation) and for assessing how well these models fit the experimental data (experimental validation). Annotation refers to the process of adding information resulting from the validation to the entry. The PDB validates structures using accepted community standards as part of ADIT's integrated data-processing system. The following checks are run and are summarized in a letter that is communicated directly to the depositor:

Covalent bond distances and angles

Proteins are compared against standard values from Engh and Huber (1991); nucleic acid bases are compared against standard values from Clowney et al. (1996); sugar and phosphates are compared against standard values from Gelbin et al. (1996).

Stereochemical validation

All chiral centers of proteins and nucleic acids are checked for correct stereochemistry.

Atom nomenclature

The nomenclature of all atoms is checked for compliance with International Union of Pure and Applied Chemistry (IUPAC) standards (IUPAC-IUB) and is adjusted if necessary.

Close contacts

The distances between all atoms within the asymmetric unit of crystal structures and the unique molecule of NMR structures are calculated. For crystal structures, contacts between symmetry-related molecules are checked as well.

Ligand and atom nomenclature

Residue and atom nomenclature are compared against a standard dictionary for all ligands as well as standard residues and bases. Unrecognized ligand groups are flagged and any discrepancies in known ligands are listed as extra or missing atoms. New ligands are added to the dictionary as they are deposited.

Sequence comparison

The sequence provided by the depositor is compared against the sequence derived from the coordinate records. This information is displayed in a table where any differences or missing residues are annotated. During the annotation process the sequence database references provided by the author are checked for accuracy. If no reference is given, a BLAST search is used to find the best match. Any conflict between the depositor's sequence and the sequence derived from the coordinate records is further resolved and annotated by comparison with other sequence databases as needed.

Distant waters

The distances between all water oxygen atoms and all polar atoms (oxygen and nitrogen) of the macromolecules, ligands, and solvent in the asymmetric unit are calculated. Distant solvent atoms are repositioned using crystallographic symmetry such that they fall within the solvation sphere of the macromolecule.

In almost all cases, serious errors detected by these checks have been corrected through annotation and correspondence with the authors. It is also possible to run these validation checks against structures before they are deposited. A validation server has been made available for this purpose. In addition to the summary report letter, the server also provides output from PROCHECK, NUCheck, and SFCHECK. A summary atlas page and molecular graphics images are also produced.

The PDB continuously reviews the validation methods used and will continue to integrate new procedures as they become available and are accepted as community standards.

Data Deposition Sites

Data are deposited to the PDB to one of three sites. Because it is critical that the final archive is kept uniform, the content and format of the final files as well as the methods used to check them must be the same. The RCSB -PDB deposition site has developed software programs for data deposition, validation, and processing, including ADIT and the Validation Server. The ADIT system, as described above, is also used to process the data deposited.

The Institute for Protein Research at Osaka University in Japan has collaborated with the PDB to establish another deposition center. All data deposited at this center (primarily depositors in Asia) are also processed by this Osaka group using the ADIT system.

The Macromolecular Structure Database group at the European Bioinformatics Institute (MSD-EBI) processes data that are submitted to them via AutoDep. After processing, the data are sent to the RCSB in PDB format for inclusion in the central archive. A common mmCIF exchange dictionary has been developed with this group, which will help ensure a higher degree of data uniformity in the archival data in the future.

The PDB has also ported its data-processing software to a stand-alone system that does not require Internet access. This system is soon to be released for use by authors who wish to check data in their home laboratories.

Data Processing Statistics

Production processing of PDB entries by the RCSB began on January 27, 1999. The median time from deposition to the completion of data processing including author interactions is less than two weeks. The number of structures with a HOLD release status remains at about 16% of all submissions; 63% are hold until publication; and 21% are released immediately after processing.

Data Uniformity

A key goal of the PDB is to make the archive as consistent and error-free as possible. As indicated above, all new depositions are reviewed carefully by annotators before release. Errors found subsequent to release by authors and PDB users are addressed as rapidly as possible. Minor errors result in revisions to the entry that are annotated within the entry; major errors lead to a superceding entry or entry withdrawal.

"*Legacy data*," that is, data submitted prior to October 1998, comply with several different PDB formats, and variation exists in how the same features are described for different structures within each format. The inconsistency of formats and nomenclature conventions makes it difficult to consistently parse these data and query across the archive. As an immediate solution to the query problem, particular records across all entries in the archive were corrected; these included citation, R factor, and resolution. These corrections were loaded into the database and thus it was possible to query on these features and obtain accurate results. However, these data were not available in the PDB files. To provide uniform data for each structure we used the software that was developed and tested for primary processing and revalidated all the data in the archive. Corrections were made to nomenclature and special attention was made to consistency of the chemical description of the macromolecule and the ligand.

The original PDB files will continue to be available as they are a historical record and have been the basis of many research projects. Software is available from the PDB to transform the mmCIF files to PDB-formatted files. In the future, these mmCIF files will form the basis of the PDB databases accessible via the Web.

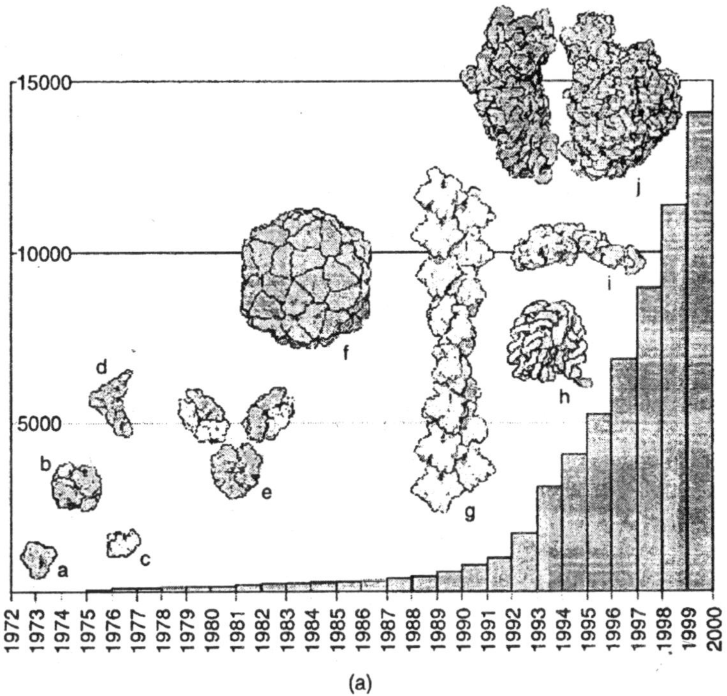

(a)

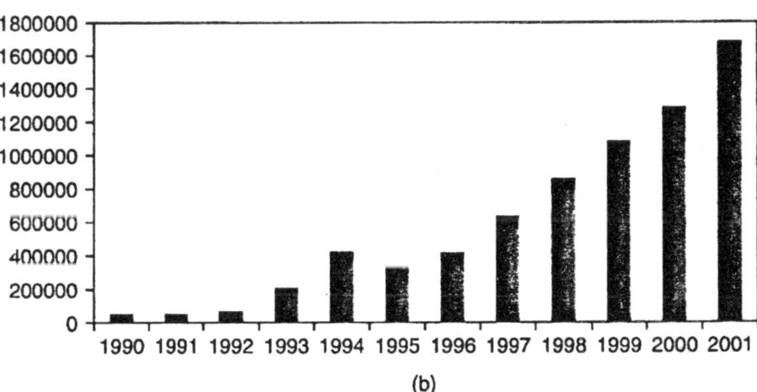

(b)

Fig. 5.2. (A) Growth chart of the PDB showing the total number of structures available in the PDB archive per year and highlighting example structures from different time periods. (B) Number of residues released in the PDB per year.

Data Access

The PDB is presently incremented once per week with new data becoming available on Wednesday mornings in most parts of the world through a number of mirror sites. The following describes the database architecture used by the PDB, how users access these databases via the Web, and how data files can be accessed via the Web and via ftp.

Database Architecture

The current PDB data management system consists of several heterogeneous data sources that are integrated through Perl CGI scripts. Although this leads to some redundancy, since parts of the data are stored multiple times, it allows efficient access. The complete system is currently being reengineered to maintain this efficiency while providing more manageability with less redundancy. The new system will be based on the DB2 relational database management system. We consider here the five core components of the current system.

First, the core relational database (Sybase SQL server release 11.0) stores the primary experimental and coordinate data. These data are retrieved by the reporting options available through the Web interface. Second, the ftp archive provides the data files in PDB and mmCIF formats as well as the data dictionaries to which they correspond. Third, the Property Object Model (POM) data management system is used for more efficient access to certain structural features, such as sequence. POM consists of indexed objects containing native data (e.g., atomic coordinates) and derived properties (e.g., secondary structure calculated according to Kabsch and Sander (1983)). Fourth, the Netscape LDAP server is used to index the textual content of the PDB and provides support for keyword searches. Fifth, the Molecular Information Agent (MIA) is used to collect and store hyperlinks and limited other information for approximately 60 external data resources in a separate Sybase database. MIA formulates a query to each of these data sources based on the PDBid, and parses the results of the query to provide the information viewable through the "Other Sources" option of an entry's Structure Explorer page. MIA includes housekeeping software, for example, to coordinate the simultaneous access to these data sources and to timeout if a particular site is down. These five components, associated software, and Web pages constitute the system that is mirrored to a number of sites worldwide.

Finally, there is a close integration to three external resources (i.e., not mirrored as part of the PDB):

1. The Biological Macromolecule Crystallization Database (BMCD) is organized as a relational database within Sybase and contains three general categories of literature derived information: macromolecular, crystal, and summary data.

2. The Nucleic Acid Database (NDB) which contains information pertaining specifically to DNA and RNA.
3. The CE database of 3D protein structure alignments.

The latter raises an important point of PDB policy. As described above, alignment of structures depends to some degree on the assumptions of the method being used. Since there is no agreement in the community at present as to a de facto standard method for protein structure alignment, the PDB's policy is to provide access to a variety of alignments and classification schemes. In short, the PDB's policy is to provide a portal (entry point) to relevant information, but to not impose judgment on which methodology should be used.

In the current implementation, communication among the five PDB components and these databases has been accomplished using the *Common Gateway Interface* (CGI) in such a way as to hide the intricacies of the underlying databases from the user. An integrated Web interface dispatches a query to the appropriate database(s), which then execute the query. Each database returns the PDBids that satisfy the query, and the CGI program integrates the results. Complex queries are performed by repeating the process and having the interface program perform the appropriate Boolean operation(s) on the collection of query results. A variety of output options are then available for use with the final list of selected structures.

The newly created and uniform mmCIFs will enable the PDB to substantially improve its underlying database architecture. The mmCIFs are loaded into a new relational database with a schema that conceptually conforms closely to the mmCIF dictionary. The results will provide access to data not currently available and do so in a way that is easier to maintain.

User Web Access

Currently, three distinct query interfaces are available for the query of data within the PDB: Status Query; SearchLite; and SearchFields.

The Status Query allows the user to review information on structures deposited but not yet released. In addition to an author list, title, and release status, the depositor may opt to release sequence information for the unreleased entry. This provides a set of useful targets for structure prediction studies.

SearchLite provides a single form field for keyword searches. Textual information within the PDB file, such as dates and some experimental data, are searched. Boolean searching and restriction of

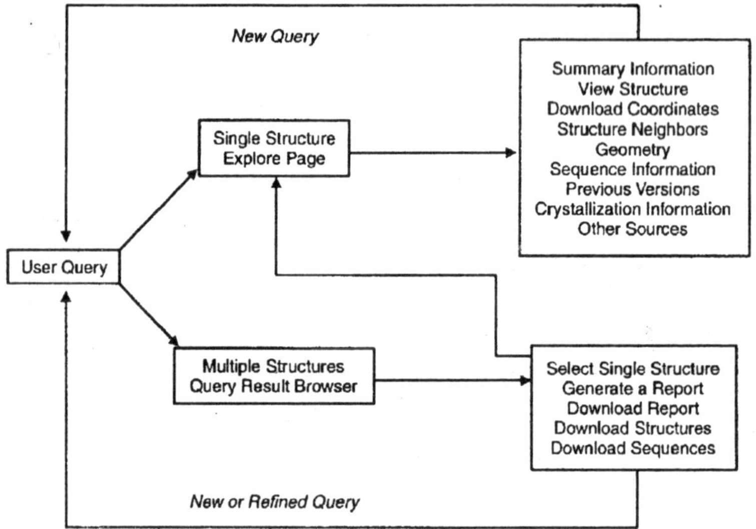

Fig. 5.3. The layout of the PDB query system.

keywords can be used to conform to specific attributes. For example, "green" can be attributed to an author name or a common name for a protein.

SearchFields is for more advanced searches and presents a customizable query form that can be used to search different data items, including macromolecule type, citation authors, sequence (via a FASTA search; Pearson and Lipman, 1988), and release dates. For enzymes it is possible to browse these structures using the Enzyme Commission hierarchy. The numbers of entries at each level are reported as the user traverses the hierarchy.

Search results are displayed in the "*Query Result Browser*," which can be used to generate reports, download data, and perform further searches. The "*Structure Explorer*" interface provides detailed information on a single structure.

Application Web Access

Interfaces to both single and multiple structures are accessible to other Web resources and applications through the simple CGI application programmer interface. Stated another way, a URL can be constructed with either single or multiple embedded PDBids and used to return results on those structures. Many Web sites worldwide use this mechanism to reference PDB structures. The PDB Web site is maintained on redundant load balanced servers and receives in excess of 100,000 page hits per day. On average a structure is downloaded every second 24 hours per day, 7 days per week.

Table 5.2. Current Query Capabilities of the PDB

Query Options	
SearchLite	Any word or combination of words in the PDB
SearchFields	*General information*: PDB identifier, citation author, chain type (protein, DNA etc.), PDB HEADER, experimental technique, deposition/release date, citation, compound information, EC number, text search
	Sequence and secondary structure: chain length, FASTA search, short sequence pattern, secondary structure content
	Crystallographic experimental information: resolution, space group, unit cell dimensions, parameters
Status	PDB identifier, deposition author, title, holding status, deposition date, release date, prereleased sequences
Result Analysis	
Single Structure: Structure Explorer	
Summary	Compound name, authors, experimental method, classification, source, primary citation, deposition date, release date, resolution, R-value, space group, unit cell parameters, polymer chain identifiers, number of residues, HET groups, number of atoms
View Structure	VRML, RasMol, QuickPDB, Chime, still images
Download/Display file	HTML and text formats for display; PDB and mmCIF formats with different compression options for download
Structural Neighbors	List of sites for finding structural homologues
Geometry	Unusual dihedral angles, bond angles and bond lengths
Other Sources	Links to other sources of information
Sequence Details	Chain Ids, number of residues per chain, molecular weight, chain type, secondary structure assignment; download sequence only in FASTA format
Crystallization Information	Conditions under which the crystals were obtained
Previous versions	Versions of the structure replaced by the current version if applicable
Nucleic Acid Database Atlas Entry	Detailed information from the NDB (if applicable)

Quick Report	Nucleic acid geometry if applicable
Structure Factors	Experimental data if available
Multiple Structure: Results Browser	
Summary List	Deposition date, resolution, experimental method, classification, compound name
Download Structures or Sequences	mmCIF and PDB compressed files (gzip, tar, compressed); sequences in FASTA format
Query Refinement	Iterative query over result set using OR, AND or NOT Boolean logic
Tabular Report	Cell dimensions, primary citation, structure identifiers, sequence, experimental details, refinement details
Query Review	Summary of queries submitted thus far with the option to return to one of them

ftp Access

All structures, in PDB and mmCIF formats, are available for download from the PDB ftp site. Dictionaries, documentation, and PDB-provided software are also available.

Distribution

As stated, the PDB distributes coordinate data in PDB and CIF formats, structure factor files, and NMR constraint files. In addition, it provides derived data, documentation, and software. New data officially become available at 2:00 A.M. Pacific standard time each Wednesday. PDB mirrors have been established in Japan, Singapore, Brazil, and in the UK. Additionally, other sources of PDB data exist, but are provided through different interfaces. The PDB also distributes a quarterly CD-ROM set that is essentially a copy of the ftp site. Data are distributed as compressed files using the compression utility program gzip. There is no charge for this service.

Outreach

Active outreach ensures that the community of PDB users is fully informed about our capabilities and activities and that the PDB receives feedback that allows it to improve its services. Outlined below are some of the key outreach activities.

Help desk

The electronic help desk addresses questions about all aspects of the PDB and about general structural biology. Questions are generally addressed within one or two working days. The list receives an average of 130 inquiries per month. The PDB also maintains two other addresses: deposit@rcsb.rutgers.edu, for questions concerning data deposition and help@rcsb.rutgers.edu for questions concerning ADIT.

pdb-l@rcsb.org

A list server at pdb-l@rcsb.org is maintained for use by the community to make announcements and conduct discussions on activities relating to the PDB. It is a PDB policy that this list be reserved for open discussion by the community and not for use by the PDB itself.

PDB web site

The Web site is updated weekly with news, recent developments, and improvements to existing documents. The site includes tutorials and user guides for query, deposition, and file formats.

PDB publications

The PDB publishes a quarterly newsletter available via e-mail or postal mail. PDF versions of the PDB newsletter dating back to September 1974 are available at this site. Flyers, tutorials, and an annual report are accessible from the PDB Web site.

Scientific meetings

PDB members attend a wide variety of meetings, presenting posters, talks, and exhibit booths. Among the meetings attended are the American Crystallographic Association's Annual Meeting, Protein Society's Symposiums, the Intelligent Systems in Molecular Biology's annual meeting, and the International Union of Crystallography's Congress and General Assembly.

FUTURE

Structural biology is a fast-evolving field that poses challenges to the collection, curation, and distribution of macromolecular structure data. Since 1999 the number of depositions has averaged approximately 50 per week. However, with the advent of a number of structure genomics initiatives worldwide this number is likely to increase. We estimate that the PDB could contain 35,000 structures by 2005. This growth presents a challenge to timely distribution while maintaining high quality. We believe our approach to information management should permit us to accommodate the anticipated large data influx. We are endeavoring to work closely with all structural genomics projects to automatically collect more data and are redesigning our database systems to provide a more scalable system. In terms of access, we have worked with the Object Management Group to define a Corba standard for macromolecular structure, which is closely aligned with the mmCIF dictionary. Eventually this will provide a fine-grained access to items of PDB data by users and their applications. This will be achieved by providing a Corba server that is currently under development.

The maintenance and further development of the PDB are community efforts. The willingness of others to share ideas, software, and data provides a depth to the resource not obtainable otherwise. It is important to acknowledge the contribution of scientists and staff at the BNL, who maintained the archive for many years. New input is constantly being sought and the PDB invites you to make comments at any time by sending electronic mail.

6

Macromolecular Determination

The desire to understand biological processes at a molecular level has led to the routine application of X-ray crystallography. However, significant time and effort usually are required to solve and complete a macromolecular crystal structure. Much of this effort is in the form of manual interpretation of complex numerical data using a diverse array of software packages, and the repeated use of interactive three-dimensional graphics. The need for extensive manual intervention leads to two major problems: significant bottlenecks that impede rapid structure solution, and the introduction of errors due to subjective interpretation of the data. These problems present a major impediment to the success of structural genomics efforts that require the whole process of structure solution to be as streamlined as possible. The automation of structure solution is thus necessary as it has the opportunity to produce minimally biased models in a short time. Recent technical advances are fundamental to achieving this automation and make high-throughput structure determination an obtainable goal.

High-throughput Structure Determination

Automation in macromolecular X-ray crystallography has been a goal for many researchers. The field of small-molecule crystallography, where atomic resolution data are routinely collected, is already highly automated. As a result, the current growth rate of the Cambridge Structural Database (CCSD) is more than 15,000 new structures per year. This growth rate is approximately 10 times the growth rate of the Protein Data Bank (PDB). Automation of macromolecular crystallography could significantly improve the rate at which new structures are determined. Recently, the goal of automation has moved to a position of prime importance with the development of the concept of structural genomics.

In order to exploit the information present in the rapidly expanding sequence databases, the structural database must also grow. Increased knowledge about the relationship between sequence, structure, and function will allow sequence information to be used to its full extent. For structural genomics to be successful, macromolecular structures will need to be solved at a rate significantly faster than at present. This high-throughput structure determination will require automation to reduce the bottlenecks related to human intervention.

Automation will rely on: the development of algorithms that minimize or eliminate subjective input; the development of algorithms that automate procedures that were traditionally performed by hand; and, finally the development of software packages that allow a tight integration between these algorithms. Truly automated structure solution will require the computer to make decisions about how best to proceed in the light of the available data.

The automation of macromolecular structure solution applies to all of the procedures involved. There have been many technological advances that make macromolecular X-ray crystallography easier. In particular, cryoprotection to extend crystal life, the availability of tunable synchrotron sources, high-speed *charge-coupled device* (CCD) data collection devices, and the ability to incorporate anomalously scattering selenium atoms into proteins have all made structure solution much more efficient. The desire to make structure solution more efficient has led to investigations into the optimal data collection strategies for multiwavelength anomalous diffraction and phasing using single anomalous diffraction with sulfur or ions.

Gonzalez and her colleagues have shown that multi-wavelength anomalous diffraction (MAD) phasing using only two wavelengths can be successful. The optimum wavelengths for such an experiment are those that give a large contrast in the real part of the anomalous scattering factor (e.g., the inflection point and high-energy remote). However, Rice and his colleagues have also shown that, in general, a single wavelength collected at the anomalous peak is sufficient to solve a macromolecular structure. Such an approach minimizes the amount of data to be collected and increases the efficiency of synchrotron beamlines, and is therefore likely to become an important and widely used technique in the future.

Data Analysis

the appropriate signal extracted. Observations that are in error must be rejected as outliers. Some observations will be rejected at the data-processing stage, where multiple observations are available. However, if redundancy is low, then probabilistic methods can be used. The prior expectation, given either by a Wilson distribution of intensities or model-based structure-factor probability distributions, is used to detect outliers. This method is able to reject strong observations that are in error, which tend to dominate the features of electron-density and Patterson maps. This method could also be extended to the rejection of outliers during the model refinement process.

When using isomorphous substitution or anomalous diffraction methods for experimental phasing the relevant information lies in the differences between the multiple observations. In the case of anomalous diffraction, these differences are often very small, being of the same order as the noise in the data. In general the anomalous differences at the peak wavelength are sufficient to locate the heavy atoms, provided that a large enough anomalous signal is observed. However, in less routine cases it can be very important to extract the maximum information from the data. One approach used in MAD phasing is to analyze the data sets to calculate FA structure factors, which correspond to the anomalously scattering substructure. Several programs are available to estimate the FA structure factors: XPREP, MADSYS and SOLVE. In another approach, a specialized procedure for the normalization of structure factor differences arising from either isomorphous or anomalous differences has been developed in order to facilitate the use of direct methods for heavy atom location.

Merohedral twinning of the diffraction data can make structure solution difficult and in some cases impossible. The twinning occurs when a crystal contains multiple diffracting domains that are related by a simple transformation such as a twofold rotation about a crystallographic axis, a phenomenon that can only occur in certain space groups. As a result the observed diffraction intensities are the sum of the intensities from the two distinctly oriented domains. Fortunately, the presence of twinning can be detected at an early stage by the statistical analysis of structure factor distributions. If the twinning is only partial, it is possible to detwin the data. Perfect twinning typically makes structure solution using experimental phasing methods difficult, but the molecular replacement method still can be successfully used.

Heavy Atom Location and Computation of Experimental Phases

The location of heavy atoms in isomorphous replacement or the location of anomalous scatterers was traditionally performed by manual inspection of Patterson maps. However, in recent years labeling techniques such as seleno–methionyl incorporation have become widely used. Such labeling techniques lead to an increase in the number of atoms to be located, rendering manual interpretation of Patterson maps extremely difficult. As a result, automated heavy atom location methods have proliferated. The programs SOLVE and CNS use Patterson-based techniques to find a starting heavy atom configuration that is then completed using difference Fourier analyses. Both Shake-and-Bake (SnB) and SHELX-D use the direct methods reciprocal-space phase refinement combined with modifications in real-space. SnB refines phases derived from randomly positioned atoms, while SHELX-D derives starting phases by automatic inspection of the Patterson map. All methods have been used with great success to solve substructures with more than 60 selenium sites. SHELX-D and SnB have been used to find up to 150 and 160 selenium sites, respectively.

After the heavy atom or anomalously scattering substructure has been located, experimental phases can be calculated and the parameters of the substructure refined. A number of modern maximum-likelihood based methods for heavy atom refinement and phasing are readily available: MLPHARE, CNS, SHARP, SOLVE. The SOLVE program has the advantage of fully integrating and automating heavy atom location, refinement, and phasing, and therefore is very easy to use. The SHARP program implements a more complex algorithm for phasing, making use of two-dimensional integration over both phases and amplitudes. This method is computationally expensive, rendering SHARP typically an order of magnitude slower than other phasing programs, but in the case of significant nonisomorphism between heavy atom derivative data sets the improvement in the phases can be worth the additional computing time.

Density Modification

Often the raw phases obtained from the experiment are not of sufficient quality to proceed with structure determination. However, there are many real space constraints, such as solvent flatness, that can be applied to electron density maps in an iterative fashion to improve initial phase estimates. This process of density modification is now routinely used to improve experimental phases prior to map interpretation and model building. However, due to the cyclic nature of the density modification process, where the original phases are combined with new phase estimates, introduction of bias is a serious problem. The γ correction was developed to reduce the bias inherent in the process, and has been applied successfully in the method of solvent-flipping.

The γ correction has been generalized to the γ perturbation method in the DM program, part of the CCP4 suite, and can be applied to any arbitrary density modification procedure, including noncrystallographic symmetry averaging and histogram matching. After bias removal, histogram matching is significantly more powerful than solvent flattening for comparable volumes of protein and solvent. More recently a reciprocal-space, maximum-likelihood formulation of the density modification process has been devised and implemented in the program RESOLVE. This method has the advantage that a likelihood function can be directly optimized with respect to the available parameters (phases and amplitudes), rather than indirectly through a weighted combination of starting parameters with those derived from flattened maps. In this way the problem of choices of weights for phase combination is avoided. The SOLVE and RESOLVE programs together provide a relatively automated way to go from experimental data to a map suitable for model building.

Molecular Replacement

The method of molecular replacement is commonly used to solve the structures for which a homologous structure is already known. As the database of known structurese expands as a result of structural genomics efforts, this technique will become more and more important. The method attempts to locate a molecule or fragments of a molecule, whose structure is known, in the unit cell of an unknown structure for which experimental data are available. In order to make the problem tractable, it has traditionally been broken down into two consecutive three-dimensional search problems: a search to determine the rotational orientation of the model followed by a search to determine the translational orientation for the rotated model. The method of Patterson Correlation (PC) refinement is often used to optimize the rotational orientation prior to the translation search, thus increasing the likelihood of finding the correct solution. With currently available programs structure solution by molecular replacement usually involves significant manual input. Recently, however, methods have been developed to automate molecular replacement. One approach has used the exhaustive application of traditional rotation and translation methods to perform a complete six-dimensional search. More recently, less time-consuming methods have been developed. The EPMR program implements an evolutionary algorithm to perform a very efficient six-dimensional search. A Monte-Carlo simulated annealing scheme is used in the program Queen of Spades to locate the positions of molecules in the asymmetric unit.

To improve the sensitivity of any molecular replacement search algorithm, maximum likelihood methods have been developed. The traditional scoring function of the search is replaced by a function that takes into account the errors in the model and the uncertainties at each stage. This approach is seen to greatly improve the chances of finding a correct solution using the traditional approach of rotation and translation searches. In addition, the method performs a statistically correct treatment of simultaneous information from multiple search models using multivariate statistical analysis. This method will allow information from different structures to be used in highly automated procedures while minimizing the risk of introducing bias. In the future molecular replacement algorithms may permit experimental data to be exhaustively tested against all known structures to determine whether a homologous structure is already present in a database, which could then be used as an aid in structure determination.

Map Interpretation

The interpretation of the initial electron density map, calculated using either experimental phasing or molecular replacement methods, is often performed in multiple stages with the final goal being the construction of an atomic model. If the interpretation cannot proceed to an atomic model, that is often an indication that the data collection must be repeated with improved crystals. Alternatively, repeating previous computational steps in data analysis or phasing may generate revised hypotheses about the crystal, such as a different space group symmetry or estimate of unit cell contents. Clearly, completely automating the process of structure solution will require that these eventualities are taken into consideration and dealt with in a rigorous manner.

The first stage of electron density map interpretation is an overall assessment of the information contained in a given map. The standard deviation of the local root-mean-square electron density can be calculated from the map. This variation is high when the electron-density map has well-defined protein and solvent regions and is low for maps calculated with random phases. Terwilliger and Berendzen also have shown that the correlation of the local root-mean-square density in adjacent regions in the unit cell can be used as a measure of the presence of distinct, contiguous solvent and macromolecular regions in an electron density map.

Currently the process of analyzing an experimental electron density map to build the atomic model is a time-consuming, subjective process and almost entirely graphics based. Sophisticated programs such as O, XtalView, QUANTA, TurboFrodo, and MAIN are commonly used for manual rebuilding. These greatly reduce the effort required to rebuild models by providing: libraries of side chain rotamers and peptide fragments and map interpretation tools and real space refinement of rebuilt fragments. However, Mowbray and her colleagues have shown that there are substantial differences in the models built manually by different people when presented with the same experimental data. The majority of time spent in completing a crystal structure is in the use of interactive graphics to manually modify the model. This manual modification is required either to correct parts of the model that are incorrectly placed or to add parts of the model that are currently missing. This process is prone to human error because of the large number of degrees of freedom of the model and the possible poor quality of regions of the electron density map.

Although interactive graphics systems for manual model building have made the process dramatically simpler, there have also been significant advances in making the process of map interpretation and model building truly automated. One route to automated analysis of the electron density map is the recognition of larger structural elements, such as α-helices and β-strands. Location of these features can often be achieved even in electron density maps of low quality using exhaustive searches in either real space or reciprocal space, the latter having a significant advantage in speed because the translation search for each orientation can be calculated using a Fast Fourier Transform. The automatic location of secondary structure elements from skeletonized electron density maps can be combined with sequence information and databases of known structures to build an initial atomic model with little or no manual intervention from the user. This method has been seen to work even at relatively low resolution (d_{min} ~ 3.0Å). However, the implementation is still graphics based and requires user input. A related approach in the program MAID also uses a skeleton generated from the electron density map as the start point for locating secondary structure elements.

Trial points are extended in space by searching for connected electron density at C_α distance (approximately 3.7Å) with standard α-helical or β-strand geometry. Real-space refinement of the fragments generated is used to improve the model. Both of these methods suffer from the limitation that they do not combine the model-building process with the generation of improved electron density maps derived from the starting phases and the partial models.

In order to completely automate the model-building process, a method has been developed that combines automated identification of potential atomic sites in the map with model refinement. An iterative procedure is used that describes the electron density map as a set of unconnected atoms from which proteinlike patterns, primarily the main-chain trace from peptide units, are extracted. From this information and knowledge of the protein sequence, a model can be automatically constructed. This powerful procedure, known as warpNtrace in ARP/wARP, can gradually build a more complete model from the initial electron density map and in many cases is capable of building the majority of the protein structure in a completely automated way. Unfortunately, this method currently has the limitation of a need for relatively high-resolution data ($d_{min} < 2.0$Å). Data that extend to this resolution are available for less than 50% of the ~16,500 X-ray structures in the PDB.

To extend the applicability of automated map interpretation to lower resolution data, work has started using pattern recognition methods. The resulting program is called TEXTAL and shows great promise for the interpretation of maps, even at a data resolution as low as 3.0Å. Data of this quality are available for approximately 95% of the structures in the PDB. We anticipate that the combination of secondary structure fragment location, the pattern matching methods of the TEXTAL program, and iteration with structure refinement for map improvement will in the future provide a general solution to the problem of model building at resolutions better than 3.5Å.

Refinement

In general the atomic model obtained by automatic or manual methods contains some errors and must be optimized to best fit the experimental data and prior chemical information. In addition, the initial model is often incomplete and refinement is carried out to generate improved phases that can then be used to compute a more accurate electron density map. However, the refinement of macromolecular structures is often difficult for several reasons. First, the data-to-parameter ratio is low, creating the danger of overfitting the diffraction data. This method results in a good agreement of the model to the experimental data even when it contains significant errors. Therefore, the apparent ratio of data to parameters is often increased by incorporation of chemical information, that is, bond length and bond angle restraints obtained from ideal values seen in high-resolution structures. Second, the initial model often has significant errors, often due to the limited quality of the experimental data, or a low level of homology between the search model and the true structure in molecular replacement. Third, local (false) minima exist in the target function. The more local minima and the deeper they are, the more likely refinement will fail. Fourth, model bias in the electron density maps complicates the process of manual rebuilding between cycles of automated refinement.

Methods have been devised to address these difficulties. Cross validation, in the form of the free *R*-value, can be used to detect overfitting. The radius of convergence of refinement can be increased by the use of stochastic optimization methods such as molecular dynamics-based simulated annealing. Most recently, improved targets for refinement of incomplete, error-containing models have been obtained using the more general maximum likelihood formulation. The resulting maximum likelihood refinement targets have been successfully combined with the powerful optimization method of simulated annealing to provide a very robust and efficient refinement scheme.

For many structures, some initial experimental phase information is available from either isomorphous heavy atom replacement or anomalous diffraction methods. These phases represent additional observations that can be incorporated in the refinement target. Tests have shown that the addition of experimental phase information greatly improves the results of refinement. We anticipate that the maximum likelihood refinement method will be extended further to incorporate multivariate statistical analysis, thus, allowing multiple models to be refined simultaneously against the experimental data without introducing bias.

The refinement methods used in macromolecular structure determination work almost exclusively in reciprocal space. However, there has been renewed interest in the use of real-space refinement algorithms that can take advantage of high quality experimental phases from anomalous diffraction experiments or noncrystallographic symmetry averaging. Tests have shown that the method can be successfully combined with the technique of simulated annealing.

The parameterization of the atomic model in refinement is of great importance. When the resolution of the experimental data is limited, then it is appropriate to use chemical constraints on bond lengths and angles. This torsion angle representation is seen to decrease overfitting and to improve the radius of convergence of refinement. If data are available to high enough resolution, additional atomic displacement parameters can be used. Macromolecular structures often show anisotropic motion, which can be resolved at a broad spectrum of levels ranging from whole domains down to individual atoms. The use of the Fast Fourier Transform to refine anisotropic parameters in the program REFMAC has greatly improved the speed with which such models can be generated and tested. The method has been shown to improve the crystallographic R-value and free R-value as well as the fit to geometric targets for data with resolution higher than 2Å.

VALIDATION

Validation of macromolecular models and their experimental data is an essential part of structure determination. Validation is important both during the structure solution process and at the time of coordinate and data deposition at the Protein Data Bank, where extensive validation criteria are also applied. In the future, the repeated application of validation criteria in automated structure solution will help avoid errors that currently occur as a result of subjective manual interpretation of data and models.

CHALLENGES TO AUTOMATION

Noncrystallographic Symmetry

It is not uncommon for macromolecules to crystallize with more than one copy in the asymmetric unit. This result leads to relationships between atoms in real space and diffraction intensities in reciprocal space. These relationships can be exploited in the structure solution process. However, the identification of noncrystallographic symmetry (NCS) is generally a manual process. A method for automatic location of proper NCS (i.e., a rotation axis) has been shown to be successful even at low resolution. A more general approach to finding NCS relationships uses skeletonization of electron density maps. A monomer envelope is calculated from the solvent mask generated by solvent flattening. The NCS relationships between monomer envelopes can then be determined using standard molecular replacement methods.

These methods could be used in the future to automate the location of NCS operators and the determination of molecular masks. In the case of experimental phasing using heavy atoms or anomalous scatterers, it is possible to locate the NCS from the sites. The RESOLVE program automates this process such that NCS averaging can be automatically performed as part of the phase improvement procedure.

Disorder

Except in the rare case of very well-ordered crystals of extremely rigid molecules, disorder of one form or another is a component of macromolecular structures. This disorder may take the form of discrete conformational substates for side chains or surface loops, or small changes in the orientation of entire molecules throughout the crystal. The degree to which this disorder can be identified and interpreted typically depends on the quality of the diffraction data. With low-to medium-resolution data, dual side chain conformations are occasionally observed. With high-resolution data (1.5Å or better) multiple side chain and main chain conformations are often seen.

The challenge for automated structure solution is the identification of the disorder and its incorporation into the atomic model without the introduction of errors as a result of misinterpreting the data. Disorder of whole molecules within the crystal, as a result of small differences in packing between neighboring unit cells, cannot be visualized in electron density maps. However, the effect on refinement statistics such as the R and free-R value can be significant because no single atomic model can fit the observed diffraction data well.

One approach to the problem is to simultaneously refine multiple models against the data. An alternative approach is the refinement of Translation-Libration-Screw (TLS) parameters for whole molecules or subdomains of molecules. This introduces only a few additional parameters to be refined while still accounting for the majority of the disorder. However, it still remains a challenge to automatically identify subdomains.

Over the last decade of the twentieth century there have been many significant advances toward automated structure determination. Programs such as SOLVE, RESOLVE, and the warpNtrace suite combine large functional blocks in an automated fashion. The program CNS provides a framework in which different algorithms can be combined and tested using a powerful scripting language. Progress toward full automation will be made in the short term by linking existing programs together using scripting languages or the World Wide Web. However, a long-term solution will require the construction of a fully integrated system that makes use of the latest advances in crystallographic algorithms and computer science.

The software that truly automates the crystallographic process will need to be intimately associated with data collection and processing. We anticipate that the next generation of automated software will permit the heavy atom location and phasing steps of structure solution to be performed in a few minutes. This speed will enable real time assessment of diffraction data as it is collected at synchrotron beamlines. Map interpretation will be significantly faster than at present, with initial analysis of the electron density taking minutes rather than the hours or days required currently.

7

STRUCTURE DATABASE

During evolution protein sequences change due to mutations in their residues and insertions and deletions of residues. These changes give rise to families of related proteins and the earliest protein family resources, based solely on sequence data, were first established in the 1970s by the pioneering work of Dayhoff. Since then many sequence databases have been established and relationships are often detected using alignment methods based on powerful dynamic programming algorithms adapted from the realm of computer science. Such methods handle the residue insertions and deletions occurring between distant evolutionary relatives very efficiently.

The structural data has always been more sparse than the sequence data due to the technical challenges of structure determination. There is currently over two orders of magnitude discrepancy between the sequence and structure resources. Thus, while the *Protein Data Bank* (PDB) contains about 16,000 structural entries, the nucleotide sequence databank at the National Centre for Biotechnology Information (NCBI) (Gen-Bank) contains over 12 million entries.

Therefore, although the first crystal structures were solved in the early 1970s, it was not until the mid-1990s that structural classifications began to emerge, primarily with Structural Classification of Proteins (SCOP), DALI, and CATH databases and data resources. Several other classifications have arisen since (for example, DDBASE, 3Dee, DaliDD) reviewed in Holm and Sander, 1994b, and Orengo, 1994. These databases use a variety of different algorithms for comparing three-dimensional (3D) structures. They also differ in methods for measuring similarity between the structures and for clustering them into fold groups or protein superfamilies and families. However, comparisons between three of the largest classifications (SCOP, DALI, CATH) recently revealed a reasonable degree of correspondence (more than 80%) between protein families generated using different protocols.

Since structure is much more highly conserved than sequence during evolution, the discovery of structural alignment algorithms and the development of structural classifications have made a significant contribution to the understanding of evolutionary mechanisms as they have enabled much more distant evolutionary relatives to be identified. Furthermore, knowledge of a protein's structure can provide important clues to the functional mechanism and biological role of the protein, for example, protein–substrate and protein–protein interactions. Because a large proportion of the structural core of the protein (often more than 50%) is conserved even in very distant relatives, structure alignments are much more accurate than sequence alignments and this situation improves the identification of conserved structural features or sequence motifs, which are often associated with protein function.

The largest structure classifications (SCOP, CATH) currently contain between 950–1400 protein superfamilies. However, these superfamilies currently map to nearly one-third of the nonredundant sequences in the GenBank sequence database (~25% on the basis of equivalent residues). Furthermore, current structure genomics initiatives, will significantly increase the number of structures determined over the next 10 years. Current estimates predict that there will be between 30,000 and 100,000 new structures before 2010. Because of the manner in which proteins are being selected for structure determination, these new structures will be predominantly from protein families currently unrepresented in the structure classifications or very distantly related to known structural families. Therefore, we may soon have structural representatives for most of the major protein families and for those of particular medical and biological interest, although some classes of structures such as transmembrane proteins may remain difficult to determine. It is also likely that methods for detecting distant relatives will improve in parallel and as a consequence of the growth in the sequence and structure databases. Links between the sequence and structure databases will also be promoted. The InterPro initiative has integrated several sequence databases (Pfam, PRINTS, PROSITE, SWISS-PROT). There are also plans to integrate the structure databases SCOP and CATH and the European Macromolecular Structure Database (EMSD) with the sequence databases InterPro and Pfam.

Structural classifications will therefore play an increasingly important role as representatives from more of the protein families are structurally determined and the mapping between structural families and genomic sequences improves. These resources will provide key data for understanding function at the molecular level. In this chapter we describe the CATH structural classification, its development, and the methods used to update and search the resource. Analysis of the classification has revealed that some protein families are very highly populated, a finding that has important implications for understanding evolutionary mechanisms.

Historical Development

The CATH domain structure database was established in 1993 when fewer than 3000 protein structures had been determined. Nearly a decade later the database has expanded considerably and contains ~13,000 protein structure entries from the PDB comprising 33,000 structural domains. CATH also contains over 200,000 sequence domains extracted from GenBank entries and assigned to one of the 1200 CATH homologous superfamilies using profile-based approaches. Since the domain was considered to be an important evolutionary unit and also because structural prediction and homology modeling methods are often more successful on a domain basis, CATH was initially established as a domain-based database. However, sequence- and structure- based relationships are also determined between multidomain proteins, and CATH now contains families and superfamilies of multidomain proteins with links to their constituent domains.

Most publicly available structure classifications are derived using sequence-based and/or structure-based protocols. These range from the completely automated approaches of DALI and the DALI Domain Database through to the largely manual approach used in compiling the SCOP database. In the CATH database semiautomated protocols are used for clustering structures both phonetically, that is, purely on the basis of structural similarity, and phylogenetically on the basis of apparent evolutionary relatedness. Any ambiguities in the assignments from automated protocols are validated manually and major bottlenecks in the classification correspond to the detection of domain boundaries and the verification of homologous relationships.

CATH is a hierarchical classification comprising four major levels. In fact CATH is an acronym for these levels: Class, Architecture, Topology, and Homology. At the top, the protein class is determined by the secondary structure composition and packing using an automated

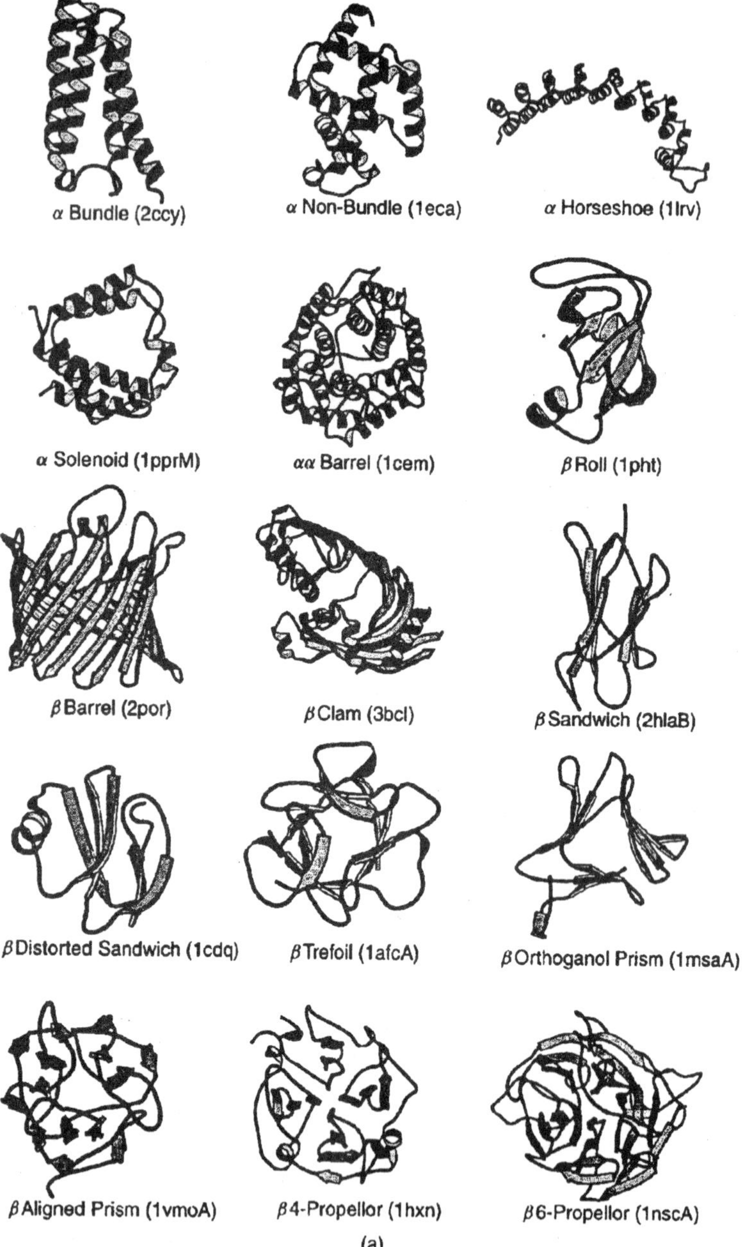

(a)

Fig. 7.1. Molscript representations of the 30 major architectures in the CATH hierarchy.

approach. Architecture describes the orientation of the secondary structures in 3D space, regardless of their connectivity. For example, a large number of protein structures adopt alpha–beta barrel architectures, in which a central barrel of beta strands is enclosed within an outer barrel comprising a layer of alpha helices. At the next level in the hierarchy, topology, both secondary structure orientation and connectivity between the secondary structures is taken into account in describing the fold of the protein.

For the example, the three-layer alpha–beta sandwich architecture contains more than 100 different folds or topologies in which the secondary structures adopt a similar shape in 3D but the connectivities between them can differ considerably as shown by the schematic representations in the illustration. At the fourth and perhaps most biologically important level in the classification, homologous superfamily, proteins are grouped according to whether there is sufficient evidence (functional similarity) to support an evolutionary relationship. Within each homologous superfamily, proteins are clustered into sequence families at different levels of sequence identity (35%, 60%, 95%, 100%). More recently, protocols have been developed for identifying functional families within each superfamily.

Table 7.1. Description of the levels in the classification at the architecture level

Primary classification number	*Description of level*
1	Mainly α
2	Mainly β
3	$\alpha\beta$
4	Few secondary structures
5	Multidomain proteins
6	Single-domain proteins classified by sequence but not structure
7	Ambiguous multidomain proteins whose domain boundary assignment requires manual validation. Protein chains clustered at the sequence level
8	New proteins classified by sequence methods
9	Chains from multichain domains classified by sequence

There are currently three major classes within CATH, corresponding to mainly- alpha domains, mainly-beta domains, and alpha-beta domains. Other categories distinguished at the class level are multidomain proteins, domains comprising few secondary structures, and three groups corresponding to proteins at different stages in the classification and pending assignment to a particular fold group or superfamily. In the December 2001 release, CATH contained 36 architectures, the major architectures, 780 fold groups, and 1390 homologous superfamilies.

Current Methodologies for Identifying Structural and Phylogenetic Relationships in CATH

Structural relationships in the CATH database were initially identified using the powerful structure-comparison algorithm, SSAP, devised by Taylor and Orengo in 1989 that incorporated a modified dynamic programming algorithm performing at two levels, thus enabling comparison of 3D information. More recently, fast graph-based methods for structure comparison have been implemented to enable the database to keep pace with the structure genomics initiatives. In fact sequence-based methods are first used to detect close relatives, as these are much faster than structure comparison and pair-wise methods are reliable for relatives with 35% or more sequence identity. Sensitive profile-based sequence methods are used to detect more distant homologues that can then be verified by structural similarity.

The strategy used in classifying new structures into the database can thus be broken down into five major steps: (1) Close relatives are identified first using pairwise sequence methods. (2) Sequence profiles and structure comparison protocols are used to detect more distant relatives. (3) Structures unclassified at this stage are then examined using both automatic and manual procedures to determine domain boundaries. (4) Unclassified domain structures are recompared using the methods employed in steps 2 and 3. (5) Finally, any structures remaining unclassified are manually assigned to architectures within CATH or new architectures. The algorithms and manual validation protocols, used at different stages of the classification, are described below.

Sequence-Based Protocols for Identifying Homologous Structures

Several studies have shown that when two proteins share more than 30% identities in their sequences, they have similar structures and can be assigned to the same superfamily. Furthermore, pairwise sequence alignment methods are reasonably robust at these levels of sequence identity. In CATH, the global alignment method of Needleman and Wunsch (1970) has been implemented and since Schneider and Sander have shown that there is a length dependence with small unrelated proteins, < 100 residues, sometimes exhibiting sequence identities as high as 30%, a more cautious threshold of 35% identity is used to detect homologues. To reduce errors further, we check that at least 80% of the larger protein is aligned against the smaller protein. Any relatives missed at this stage of the classification are captured later using structure-comparison methods.

Relatives identified by these pairwise methods are clustered into their respective families using a single linkage clustering algorithm. Single linkage was chosen because the structural data is quite sparse in some families and it is known that sequences can diverge considerably, so it is not reasonable to expect a new relative to share significant sequence similarity with a large number of other relatives in the family.

Profile-based methods are used to detect more distant homologues. A number of protocols have been implemented. PSI-BLAST and the related IMPALA algorithms developed by the Altschul group (1997) have been shown to be among the most sensitive methods available. In these approaches, a sequence is scanned against a large sequence databank (i.e., GenBank at the NCBI) initially using pairwise methods to find close homologues, which are then multiply aligned to derive a sequence profile capturing the most specific residue preferences of that sequence and its relatives. Further iterations can result in highly specific profiles for the family to which the sequence belongs. Hidden Markov Models can also be applied and the SAMT method of Karplus and co-workers has been used to build profiles for each non-identical representative in CATH and also for structurally and functionally coherent families within the database. New structures are therefore scanned against both the IMPALA and SAMT profiles for representative structures already classified in CATH. For example, there are currently 7345 SAMT profiles including those built from individual single and multidomain proteins and those generated from multiple alignments of functionally related proteins.

Both methods (IMPALA and SAMT) have been benchmarked using known relatives from CATH and parameters and thresholds optimized to give the maximum coverage for a small (< 1%) error rate. Any relatives detected using these thresholds are subsequently validated by structure comparison. The power of these sequence based approaches has been increased considerably by expanding CATH with sequence relatives from GenBank. These have been cautiously integrated into CATH superfamilies giving a 10-fold increase in the size of the database and effectively providing an intermediate sequence library for each family that broadens the scope of that family in sequence space, considerably enhancing the detection of distant homologues. Benchmarking trials using CATH demonstrated that although nearly 51% of distant structural homologues could be recognized by scanning against profiles derived from CATH structures, this percentage increased to 82% when profiles from intermediate sequence libraries for CATH were scanned.

Nearly three-quarters of all new structures are currently recognized using sequence-based methods, and the profile-based methods enable a further 10%, comprising very distant homologues, to be detected. This percentage will increase as the sequence databases and therefore the intermediate sequence libraries within CATH continue to grow with the international genome projects.

Structure-Based Methods for Identifying Structural Homologues and Related Folds (SSAP and GRATH)

A significant proportion of distant relatives, currently about 15%, can only be recognized by comparing the structures directly. There are now many examples of evolutionary relatives that have diverged to an extent where no significant sequence similarity can be detected and yet the structural fold adopted by the protein remains highly similar. For example in the globins, sequence identities of relatives can fall below 10% but the structures and oxygen-binding functions of the proteins remain highly conserved. As with sequence alignment, any method for comparing distant structural relatives must be able to cope with the extensive insertions or deletions occurring during evolution. Insertions and deletions are usually restricted to the loops between secondary structures where they are less likely to affect the fold and therefore the stability of the protein. However, in some families we have observed that secondary structures and sometimes quite large supersecondary motifs can be inserted or deleted.

In addition to residue insertions and deletions, shifts in the secondary structures can also occur to modulate the effects of volume changes caused by residue substitutions. Chothia and Lesk demonstrated quite significant movements of up to 40 degrees in some secondary structure pairs for two large protein families studied (the globins and the immunoglobulins). Analysis of families in CATH revealed that on average at least 50% of the secondary structures in the core of the protein are structurally well conserved across a family. However, although the fold of a protein family, comprising the core structural motif, may be well conserved currently, in about 10% of families there can be significant structural embellishments to this core and shifts in the orientations of the secondary structures. Sometimes these evolutionary changes give rise to modified or diverse functions within the family.

Sequential structure alignment program (SSAP)

In order to address these problems, Taylor and Orengo adapted the dynamic programming methods used so successfully in sequence alignment to compare 3D structures. Instead of comparing residue identities, the method compares the structural environments of residues between two proteins. Structural environments can be simply encoded as the set of vectors from the C- beta atom of a particular residue to the C-beta atoms of all other residues within the same protein. Since there is no prior knowledge of which residues are equivalent between the proteins, dynamic programming must be applied at two levels: a lower level in which the structural environments of all residue pairs between the proteins are compared and an upper summary level in which information from putative equivalent pairs are accumulated. The method is therefore sometimes described as double dynamic programming.

SSAP has been benchmarked and optimized using sets of validated structural homologues. A logarithmic scoring scheme was implemented that was optimized to be largely independent of the size and class of the proteins, although proteins with large proportions of alpha helices tend to give slightly higher scores as the local similarity is so highly conserved. The score is normalized to be in the range of 0 to 100 for identical proteins, irrespective of size and structures, with similar folds tending to give scores above 70, while homologous proteins often give higher scores of 80 and above. When classifying new structures, scores of 70 and above are required before proteins are assigned to a particular fold group and again single-linkage clustering is used as some families are too small to ensure that a significant proportion of known relatives will have clear structural similarity (i.e., SSAP >70) to the new relative. Although high SSAP scores suggest that the proteins may be homologues, these proteins are not assigned at the superfamily level unless there is other evidence to support homology; for example, sequence similarity detectable by PSI-BLAST or functional similarity.

Proteins are only classified into existing families of fold groups if the structural similarity detected extends for a significant proportion (more than 60%) of the larger structure. This preserves the domain-based nature of the classification. Thus, although new multidomain structures may contain one or more domains that match existing CATH families, these are not clustered into their families until a later stage in the classification when their domain boundaries have been reliably identified.

Graphical method for identifying folds (GRATH)

Although SSAP has proven to be reliable and accurate, it is computationally expensive and large structures such as TIM barrels with more than 300 residues can take several days to scan against the database, using the most powerful machines currently available. This has not proved to be a major bottleneck to date; however, structure genomics initiatives are expected to increase substantially the numbers of structures determined annually. Currently 50 to 100 new structures are determined weekly and this number may double or treble over the next decade. Therefore, to keep pace with structure genomics, a fast prefilter for SSAP has been designed, based on graph theory. Similar approaches have previously been implemented in the algorithm POSSUM designed by Artymiuk and co-workers.

The method implemented in CATH–GRATH compares secondary structures between proteins, as there are an order of magnitude fewer of these than residues. These are represented as linear vectors and are associated with nodes in a protein graph. Edges between the nodes are characterized by the orientations of the secondary structures and the distances between their midpoints. Additional angles are used to describe the tilt and rotations of the vectors. Ullmanns subgraph isomorphism algorithm is used to detect corresponding structural motifs between proteins from comparison of their graphs. Parameters for recognizing fold similarities have been optimized using validated relatives from CATH. GRATH recognizes the correct fold within the top 10 matches of a database search 98% of the time. Of importance, GRATH is 1000 times faster than SSAP for most proteins and the top 10 matches returned from a search can be validated using the more accurate SSAP method. Robust statistics (expectation values or E-values) have also been developed based on the extreme value distribution observed for a typical database scan. These statistics indicate the significance associated with a match and are important for determining the order in locating individual domain folds within multidomain proteins. Because GRATH is so fast, multidomain proteins can now be structurally compared to detect potential homologies that are then validated manually.

Methods for Generating Multiple Structure Alignments (CORA) and Protocols for Using 3D-Templates to Identify Distant Structural Relationships

Multiple structure alignments are generated for each superfamily in CATH using a modified version of the SSAP algorithm (consensus residue attributes, CORA). This algorithm is based on progressive alignment of relatives using a single-linkage tree derived from the pairwise SSAP similarity scores. The most similar structures are aligned first and the next relative, most similar to the aligned structures, is then iteratively selected and aligned in the order dictated by the tree. After addition of each relative, a consensus structure is derived consisting of average vectors between residue positions and information on the variability of these vectors. Further relatives are aligned against this consensus structure, weighting the alignment of conserved positions more highly.

Once all relatives have been aligned, information on the conservation of the residue structural environment, including residue contacts and various other attributes (e.g., accessibility, torsional angles), is compiled and encoded in a 3D template representing the set of structures. In a manner analogous to the improvements obtained using sequence profiles, structural templates have been shown to be far more effective at recognizing distant homologues as they capture the most conserved structural characteristics of the family or superfamily. New structures can be scanned against libraries of CORA templates, again using the double dynamic programming algorithm and the percentage of highly conserved contacts for the superfamily that are present in the putative relative is calculated (ConA), to determine homology. The pattern of conserved residue contacts has been shown to be a highly characteristic topological fingerprint for a particular superfamily.

CORA templates have been generated for each superfamily containing two or more nonidentical structures, and in the more highly populated superfamilies there are multiple templates corresponding to clusters of more closely related structures or families within the superfamily. Scanning against the template library is up to 100 times faster than performing pairwise searches with SSAP on representative structures from the superfamilies. The templates also show increased sensitivity and selectivity in recognising structural relatives over the pairwise SSAP and these attributes can be expected to increase as more in recognising structural relatives structures are determined and coverage within each superfamily increases.

Identification of Domain Boundaries

Any proteins unclassified by the sequence-based methods are divided into their constituent domains where relevant, before resubmitting them to the sequence searches and to the slower structure-comparison protocols, which often facilitates the structure alignment as a smaller region of the comparison matrix is searched. Identification of domain boundaries is a difficult process and although numerous automatic algorithms have been devised, most have an error rate of between 20% and 30% associated with them. The problem arises from the fact that no simple quantitative definition of a structural domain exists. Qualitatively, domains have been described by various authors as compact semi-independent folding units and many algorithms attempt to locate them by searching for large hydrophobic clusters indicative of domain cores and also by dividing structures so as to maximize internal residue contacts within putative domains and to minimize external contacts between them.

In the October 2001 release of CATH, nearly 40% of PDB entries are multidomain proteins, of which two-thirds comprise only two domains. Furthermore nearly one-quarter of domains in CATH are discontiguous; that is, the domain is formed from discontiguous segments of the polypeptide chain, and these cases are particularly difficult for the algorithms to detect. To improve accuracy, we employ a consensus-based protocol DBS (*Domain Boundary Suite*) that applies three independent algorithms PUU, Domak and DETECTIVE. Where they agree within a tolerance of 10 residues, domains can be assigned completely automatically. Otherwise each of the individual assignments is manually checked.

In addition we use a more recently developed protocol, employing GRATH and based on the concept that domains are known to recur in different multidomain contexts. It is now well established that domain shuffling is a common evolutionary mechanism, often responsible for creating new or modified functions within an organism. Recent analysis of CATH revealed that at least 70% of the domains from multidomain proteins recurred in different multidomain families or also occurred as single domain proteins. Therefore, a simple domain-detection protocol is now used to search for known domains within new multidomain structures. This protocol uses GRATH to compare the secondary structure graph for each new putative multidomain protein with the library of graphs from representatives from all the CATH domain families. Domains found within the multidomain protein can then be extracted in order of decreasing statistical significance. This approach has led to significant improvements in domain assignment as the E-values returned by GRATH give a measure of the reliability in boundary prediction. Any ambiguous assignments can be manually validated by coloring putative domains in the excellent molecular viewing package, RASMOL. The majority of errors are found in this way, although in difficult cases boundary identification is a somewhat subjective process as witnessed by the fact that about 17% of boundary definitions in the SCOP and CATH databases disagree. This step is one of the most time-consuming but important stages in the classification. Although domain-boundary detection from structural data is difficult, it is considerably easier than assignment from sequence data and many sequence databases now incorporate the SCOP and CATH boundaries or use them to validate their own assignments.

Structural and Functional Validation of Homologues

Sequence, structural, and functional data for each homologous superfamily in CATH are stored within a *Dictionary of Homologous Superfamilies* (DHS), which is also accessible over the Web. This dictionary was originally established in 1999 by Bray and now also contains all the sequence relatives for CATH superfamilies identified in GenBank. Information on pairwise relationships between all non-identical structures and sequences is also stored, namely, sequence identities, expectation values from PSI-BLAST and SAM-T99; SSAP structural similarity scores and expectation values from GRATH, where appropriate.

Pairwise SSAP alignments between all nonidentical structures in the superfamily are also stored. Plots of sequence identities versus structural similarity scores can be used to illustrate the structural plasticity observed within a given superfamily, and any obvious outliers may indicate problems in the alignment. CORA multiple-structure alignments are generated for each superfamily and can be viewed over the Web. These alignments are annotated in various ways: by residue identities or physicochemical properties; by secondary structure; and by PROSITE motifs. RASMOL viewers for multiple superpositions of relatives allow the 3D location of conserved sequence motifs to be easily identified.

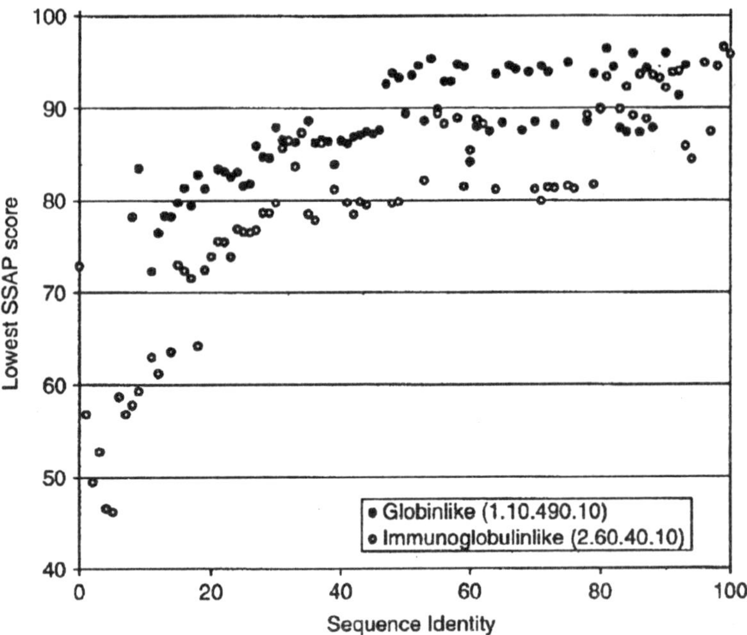

Fig. 7.2. Structural plasticity plots showing the structural similarity for pairs of relatives in selected CATH superfamilies, as measured by the SSAP algorithm as a function of sequence identity measured after structural alignment.

Of importance, the DHS contains increasing amounts of functional data for each superfamily. These data are extracted from other public resources, for example, PDB header records, PDBsum; SWISS-PROT; the Enzyme database; GenProtEC, and PROSITE. In future releases the DHS will also provide functional annotations collected from InterPro, Pfam, Gene Ontology, and metabolic pathway data. Summary tables in the DHS can be used to peruse functional attributes for relatives across the family. This information can be used diagnostically to determine whether a new structural relative possesses a similar function to other members of the family and is therefore clearly a homologue.

However, since recent analyses of CATH have revealed the extent to which function can be modified in some protein families, the DHS also contains all the sequence relatives detected for each superfamily, which considerably expands the functional information available for a particular superfamily, making it easier to find a relative with related function. It also reveals the extent to which functional change can occur within the family and this can be taken into account when assessing whether a protein with a similar fold is in fact a homologue. Protocols are currently being devised for automatically comparing functional annotations between proteins in order to speed up the validation of homologues during classification. However, we do not expect this approach to be viable for more than about 30% to 50% of new relatives, a result suggested by the performance of similar protocols employed by other groups. However, because this validation stage is time consuming and another major bottleneck in the classification, we expect the new protocols to significantly improve the frequency of CATH releases.

Recruiting Sequence Relatives into CATH Superfamilies

In order to gain as broad an understanding of function as possible within each protein family and superfamily, sequence relatives are regularly recruited into CATH from GenBank. This recruitment is achieved using the profile-based methods described above, namely, the 1D-profiles (PSI-BLAST) and iterative Hidden Markov Models (SAMT). Nonidentical representatives from each class within CATH are scanned against Gen-Bank using both PSI-BLAST and SAM-T99. A consensus method, DomainFinder, which uses conservative thresholds established by benchmarking against validated structural homologues, is then used to extract those domain regions from GenBank entries that can safely be assigned to CATH superfamilies. As with structural entries, pairwise sequence alignment methods are subsequently used to update sequence relationships within the superfamily and to enable clustering into sequence families at 35%, 60%, 95%, and 100% identities. CATH currently contains nearly 200,000 sequence relatives identified in this way.

GENE3D RESOURCE

A related CATH-based Web resource, Gene3D, uses these assignments of gene sequences to CATH superfamilies to provide structural annotations for completed genomes. The June 2001 release of Gene3D contains 36 complete genomes obtained from GenBank. For each gene within the genome, the location of gene regions that match individual structural domains from CATH superfamilies are shown, together with a consensus domain region determined by the DomainFinder algorithm. Links to the relevant CATH superfamily and DHS entries are also provided. Statistics are given on the distributions of superfamilies and fold groups identified within each genome, enabling comparison of fold and superfamily usage between genomes.

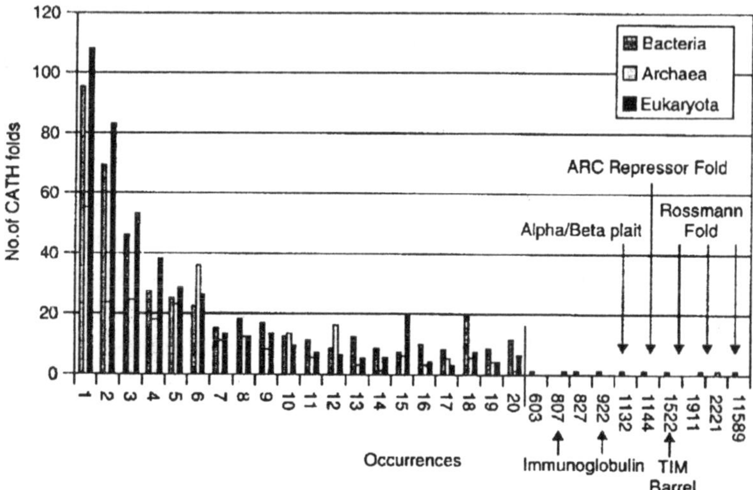

Fig. 7.3. Comparison of fold usage in the eukaryotic, archaeal, and bacterial kingdoms using the November 2000 release of CATH.

CATH Web Site and Server

The site can be used for browsing the hierarchy, and there are representative MOLSCRIPT illustrations at each level together with links to other local resources (DHS, Gene3D, PDBsum). Each level has its own unique numeric identifier that is never changed, although some numbers may disappear; for example, if new evidence suggests two superfamilies should be merged.

CATH data are stored within an Oracle 8i relational database. Special Oracle features are utilized for storing relationships and family membership and another Oracle feature, materialized views, is used to create specialized views of the data for efficient searching. The database also contains genome information used by the Gene3D resource. The schema for CATH is part of a much larger schema for a protein family database (PFDB), which also contains sequence-based families identified using other classification protocols and currently contains data for virus families, eye proteins, and other biological families being studied locally.

The database has a flexible design that allows proteins to participate in different structural and functional relationships, that is, alternative methods of clustering can easily be accommodated. The flexible design also has the benefit of allowing mapping between families or clusters generated using different protocols or based on different underlying philosophies, for example, functional versus structural.

Regular Updates of the CATH Web Site

The protocols used for updating CATH have already been described above. To keep pace with structure genomics initiatives, sequence-based classification of newly determined structures will be run at weekly intervals and clearly identified relatives recruited into the database. Monthly releases of the database are planned to make this data publicly available on a more frequent basis. The assignment of domain boundaries is more time consuming as some validation is required where GRATH and DBS results are ambiguous. Structure classification is also slower as validation is required to confirm homologues. Therefore, both domain-boundary identification and structural classification are ongoing processes and those structures whose boundaries have been determined and classified by sequence- and structure-based methods will be added to the database to coincide with monthly updates. In future, we plan to display all newly determined protein structures that meet the criteria for integration in CATH (i.e., well resolved and not model or synthetic proteins) using additional protein classes to reflect the extent to which the structures have been classified in the database.

CATH Server

Newly determined structures can be submitted to the CATH server, which scans them against representatives from the database in order to determine the putative superfamily or fold group to which the structure belongs or whether the protein comprises one or more novel folds. Domain boundaries can be supplied by the user or alternatively will be determined automatically using the consensus approach. GRATH is first used to identify a probable fold group. Subsequently, the new domain structure(s) is scanned against 3D templates for all the homologous superfamilies within the top three folds matched to identify the superfamily. Lists of structural neighbors are provided together with links to the appropriate superfamilies in CATH and the DHS. The user can also view superpositions of the structure with other relatives from the superfamily or fold group.

Statistics on the Populations of Different Levels in the CATH Hierarchy

Population of folds within architectures

Although there are currently 36 architectural groups in CATH only 28 can be well defined, and the remaining groups can be thought of as bins comprising assorted irregular or complex folds or folds containing few secondary structures and often stabilized by a high proportion of disulphide bridges. Furthermore, among the well-defined architectures, some are much more highly populated than others with about half of the folds adopting one of six regular symmetric architectures; the mainly-α bundles, the two-layer β-sandwiches and β-barrels, and the two-layer and three-layer $\alpha\beta$-sandwiches and the $\alpha\beta$-barrels. Recent analysis of structural relationships between fold groups in CATH using the GRATH algorithm has revealed that some fold groups are particularly "gregarious," that is, they have large motifs in common with many other folds within the database. For example, a high proportion of folds within the mainly-β and the $\alpha\beta$-sandwich architectures can be categorized as gregarious in this way. This idea conforms with the idea of a structural continuum between fold groups, first suggested by an early analysis of CATH. However, the recent analysis also clearly showed that some 15% of folds, for example the beta trefoil interleukin fold, are much more distinct, representative more of discrete islands than progressions in a structural continuum.

The idea of a structural continuum is perhaps not surprising as it has long been known that certain structural motifs are favored and often recur within folds for a particular class; for example, the mainly-β hairpins, the beta Greek keys, and the $\alpha\beta$ and split $\alpha\beta$ motifs. The TIM barrel fold in the $\alpha\beta$ class comprises eight recurring $\alpha\beta$ motifs, which are also found to recur in the $\alpha\beta$ Rossmann folds, though the connectivity between two of the motifs differs, giving rise to a different topology for the fold.

Population of superfamilies and families within folds

The population of the different levels in the CATH hierarchy, for the November 2000 release, that the number of new folds being determined each year is slowly decreasing. Early analysis of CATH revealed a small number of fold groups (~10) that were very highly populated containing many different homologous superfamilies and families. Re-examining these data six years later, following a nearly 10-fold expansion of the database from 3000 to 33,000 domain structures, it is clear that the trend still applies and it can be seen that 5 large fold groups contain nearly one-fifth of all the homologous superfamilies in CATH. These highly populated fold groups have been described as superfolds and other classifications (e.g., SCOP) have reported similar observations of frequently occurring domains within the database.

The popularity of these folds may be a result of divergent or convergent evolution. Divergent evolution gives rise to families of

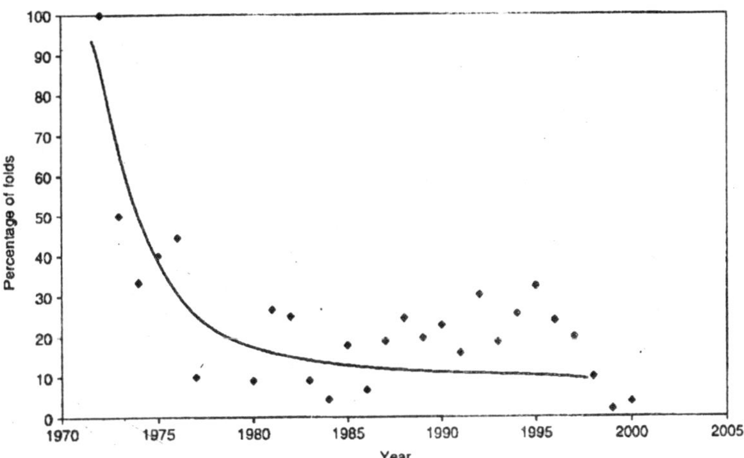

Fig. 7.4. Number of unique folds identified annually as a percentage of the number of new structures determined.

proteins in which the structure is generally well conserved but sequences may have changed to the extent that no significant similarity remains. In paralogues, which arise from duplication of the gene within an organism, the function of protein may also have been modified or changed. Thus, apparently diverse superfamilies within these superfolds may in fact be extremely distant relatives whose relationships cannot easily be verified from the available sequence or functional data.

Alternatively, Ptitsyn, Finkelstein, and others have suggested that there may be a limited number of folds in nature due to physical constraints on the packing of secondary structures. Chothia (1993) has suggested approximately 1000 folds and other similar estimates of a few thousand folds have also been made. Structures sharing the same fold but arising from different ancestral proteins have been described as analogues.

The task of determining homology is often complicated by the lack of evolutionary clues. Murzin and others have shown several interesting examples of relatives possessing similar folds but disparate sequences and functions but where homology is suggested by the presence of some rare structural characteristic, for example, a conserved β-bulge. Detailed knowledge of the structural family can often provide insights that promote detection of evolutionary fingerprints. However, such practices are not readily amenable to automated protocols. The TIM barrel fold is one of the most highly populated, with 18 superfamilies in a recent CATH release. However, several detailed studies recently have provided evidence of common ligand-binding motifs or unusual structural characteristics, suggesting that many of these superfamilies should be merged. In this context, an analysis of functional properties observed within 167 enzyme superfamilies in CATH revealed that when all known relatives for the superfamily were considered, including sequence relatives identified in the genomes, more than 25% of the families exhibited functional diversity.

Structural Quality Assurance

The experimentally determined three-dimensional (3D) structures of proteins and nucleic acids represent the knowledge base from which so much understanding of biological processes has been derived over the last three decades of the twentieth century. Individual structures have provided explanations of specific biochemical functions and mechanisms, while comparisons of structures have given insights into general principles governing these complex molecules, the interactions they make, and their biological roles.

The 3D structures form the foundation of structural bioinformatics; all structural analyses depend on them and would be impossible without them. Therefore, it is crucial to bear in mind two important truths about these structures, both of which result from the fact that they have been determined experimentally. The first is that the result of any experiment is merely a *model* that aims to give as good an explanation for the experimental data as possible. The term *structure* is commonly used, but you should realize that this should be correctly read as *model*. As such the model may be an accurate and meaningful representation of the molecule, or it may be a poor one. The quality of the data and the care with which the experiment has been performed will determine which it is. Independently performed experiments can arrive at very similar models of the same molecule; this suggests that both are accurate representations, that they are good models.

The second important truth is that any experiment, however carefully performed, will have errors associated with it. These errors come in two distinct varieties: systematic and random. Systematic errors relate to the *accuracy* of the model—how well it corresponds to the true structure of the molecule in question. These often include errors of interpretation. In X-ray crystallography, for example, the molecule(s) need to be fitted to the electron density computed from the diffraction data. If the data are poor and the quality of the electron density map is low, it can be difficult to find the correct tracing of the molecule(s) through it. A degree of subjectivity is involved and errors of mistracing and frame-shift errors, described later, are not uncommon. In NMR spectroscopy, judgments must be made at the stage of spectral interpretation where the individual NMR signals are assigned to the atoms in the structure most likely to be responsible for them.

Random errors, on the other hand, depend on how precisely a given measurement can be made. All measurements contain errors at some degree of precision. If a model is essentially correct, the sizes of the random errors will determine how *precise* the model is. The distinction between accuracy and precision is an important one. It is of little use having a very precisely defined model if it is completely inaccurate.

The sizes of the systematic and random errors may limit the types of questions a given model can answer about the given biomolecule. If the model is essentially correct, but the data was of such poor quality that its level of precision is low, then it may be of use for studies of large scale properties—such as protein folds—but worthless for detailed studies requiring the atomic position to be precisely known; for example, to help understand a catalytic mechanism.

Structures as Models

To make the point about 3D structures being merely models it is instructive to consider the subtly different types of model obtained by the two principal experimental techniques: X-ray crystallography and NMR spectroscopy. The models are of the protein rubredoxin with a bound zinc ion held in place by four cysteines.

Models from X-ray crystallography

It is not a standard depiction of a protein structure; rather, its aim is to illustrate some of the components that go into the model. The components are: the *x*-, *y*-, *z*- coordinates, the *B*-factors, and occupancies of all the individual atoms in the structure. These parameters, together with the theory that explains how X-rays are scattered by the electron clouds of atoms, aim to account for the observed diffraction pattern. The *x*-, *y*-, *z*-coordinates define the mean position of each atom, whereas its *B*-factor and occupancy aim to model its apparent disorder about that mean. This disorder may be the result of variations in the atom's position in time, due to the dynamic motions of the molecule, or variations in space, corresponding to differences in conformation from one location in the crystal to another, or both. The higher the atom's disorder, the more "smeared out" its electron density. *B*-factors model this apparent smearing around the atom's mean location; at high resolution a better fit to the observations can often be obtained by assuming the *B*-factors to be anisotropic. Occasionally, the data can be explained better by assuming that certain atoms can be in more than one place, due, say, to alternative conformations of a particular side chain. The atom's occupancy defines how often it is found in one conformation and how often in another.

Models from NMR spectroscopy

The data obtained from NMR experiments are very different, so the models obtained differ in their nature, too. The spectra measured by NMR provide a diversity of information on the average structure of the molecule, and its dynamics, in solution. The most numerous, but often least precise, data are from Nuclear Overhauser Effect SpectroscopY (NOESY) experiments where the intensities of particular signals correspond to the separations between spatially close protons (≤6Å) in the structure. The spectra from COrrelated SpectroscopY (COSY)-type experiments give more precise information on the separations of protons up to three covalent bonds apart, and in some cases on the presence, or even length, of specific hydrogen bonds. Recently developed dipolar-coupling experiments give information on the relative orientation of particular backbone covalent bonds.

For the vast majority of NMR experiments, the sample of protein or nucleic acid is in solution, rather than in crystal form, which means that molecules that are difficult to crystallize, and hence impossible to solve by crystallography, can often be solved by NMR instead. The separations are converted into distance and angular restraints and models of the structure that are consistent with these restraints are generated using various techniques, most commonly molecular dynamics-based simulated annealing procedures similar to those used in X-ray structure refinement. The end result is not a single model, but rather an ensemble of models that are all consistent with the given restraints.

The reasons for generating an ensemble of structures from NMR data are twofold. Firstly, the NMR data are relatively less precise and less numerous than experimental restraints from X rays so that a diversity of structures are consistent with them. Secondly, the biomolecules may genuinely possess heterogeneity in solution.

For general use, an ensemble of models is rather more difficult to handle than a single model. Ensembles deposited in the Protein Data Bank (PDB) can typically comprise 20 models. One of these is often designated as representative of the ensemble, or a separate file containing an average model may be deposited in addition to the ensemble. The separate average structure is energy minimized to counteract the unphysical bond lengths and angles that the averaging process introduces. Such a structure tends to have a separate PDB identifier from that of the ensemble—so the same structure, or rather the outcome of the same experiment, appears as two separate entries in the PDB. This is clearly potentially confusing and the use of separate files is now discouraged. The representative member of an ensemble is usually taken to be the structure that differs least from all other structures in the ensemble. An algorithmic Web-based tool called OLDERADO allows you to select such a representative from an ensemble, but no single algorithm is universally agreed upon.

Aim

The aim of this section is to demonstrate that not all structures are of equally high quality, usually because of the quality of the experimental data from which they were determined, and that care needs to be taken before using any structure to draw biological or other conclusions. When selecting data sets for deriving general principles about, say, protein structures it is important to filter out those that might give misleading results simply because they are unlikely to be sufficiently accurate or precise to contribute meaningful or correct data to the analysis.

It does seem slightly churlish to reject structures from consideration given the amount of time, care, and hard work the experimentalists have put into solving them. However, if you put unsound data into your analysis, you will get unsound conclusions out. This chapter hopes to explain the limitations of using 3D structures uncritically for structural bioinformatics purposes, and to provide some rules of thumb for weeding out the defective ones: what are the symptoms, what should you look for, and which structures should you reject?

Error Estimation and Precision

All scientific measurements contain errors. No measurement can be made infinitely precisely; so, at some point, say after so many decimal places, the value quoted becomes unreliable. Scientists acknowledge this by estimating and quoting standard uncertainties on their results. For example, the latest value for Boltzmann's constant is $1.3806503(24) \times 10^{-23}$ JK^{-1}, where the two digits in brackets represent the standard uncertainty in the last two digits quoted for the constant.

Compare this with the situation we have in relation to the 3D structures of biological macromolecules. It relates to a single amino acid residue (a lysine) and shows the information deposited about each

	Atom number	Atom name	Residue name	Residue number	Atomic coordinates x	y	z	Occupancy	B-factor
ATOM	1	N	LEU	1	-15.159	11.595	27.068	1.00	18.46
ATOM	2	CA	LEU	1	-14.294	10.672	26.323	1.00	9.92
ATOM	3	C	LEU	1	-14.694	9.210	26.499	1.00	12.20
ATOM	4	O	LEU	1	-14.350	8.577	27.502	1.00	13.43
ATOM	5	CB	LEU	1	-12.829	10.836	26.772	1.00	13.48
ATOM	6	CG	LEU	1	-11.745	10.348	25.834	1.00	15.93
ATOM	7	CD1	LEU	1	-11.895	11.027	24.495	1.00	13.12
ATOM	8	CD2	LEU	1	-10.378	10.636	26.402	1.00	15.12

Fig. 7.5. An extract from a PDB file of a protein structure showing how the atomic coordinates and other information on each atom are deposited.

atom in the protein's structure. Looking at only the columns representing the *x*-, *y*-, *z*-coordinates you will notice that each value is quoted to three decimal places. This suggests a precision of 1 in 10^5. Similarly, the *B*-factors (in the final column) are each quoted to two decimal places. Is it possible that the atomic positions and *B*-factors were really so precisely defined? What are the error bounds on these values? What are their *s.u.*s? Are the values accurate to the first place of decimals? The second? The third?

In fact, with the exception of a very few PDB structures, no error bounds are given. As at November 2001 there were 5 such exceptions out of 16,646 structures: one was a carbohydrate, three were marginally differing copies of the same 13-residue enterotoxin, and the fifth was the crystal structure of the 54-residue rubredoxin (4rxn). All had been solved at atomic resolution (ranging from 0.89Å to 1.2Å) and refined by the full-matrix least-squares method that is mentioned below.

Thus, in the overwhelming majority of cases one cannot tell how precisely defined the values are. Why is this so? What kind of scientific measurement is this? And how are we to judge how much reliance to place on the data given?

Error Estimates in X-Ray Crystallography

Estimation of standard uncertainties

In X-ray crystallography it is, in theory, possible to calculate the standard uncertainties of the atomic coordinates and *B*-factors. In fact, it is routinely done for the crystal structures of small molecules such as those deposited in the Cambridge Structural Database (CSD). The calculations of the *s.u.*s are performed during the refinement stage of the structure determination. Refinement involves modifying the initial model to improve the match between the experimentally determined structure factors—as obtained from the observed X-ray diffraction

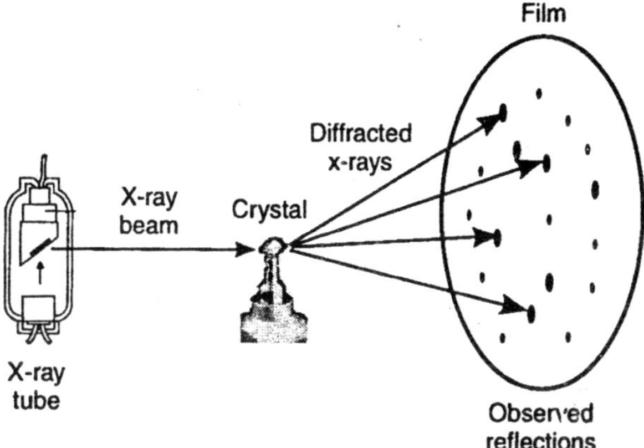

Fig. 7.6. A schematic diagram illustrating the principle of structure refinement in X-ray crystallography.

pattern—and the calculated structure factors—as obtained from applying scattering theory to the current model of the structure.

In practice refinement is usually a long drawn-out procedure requiring many cycles of computation interspersed here and there with manual adjustments of the model using molecular graphics programs to nudge the refinement process out of any local minimum that it may have become trapped in. Furthermore, because in protein crystallography the data-to-parameter ratio is poor (the data being the reflections observed in the diffraction pattern and the parameters being those defining the model of the protein structure: the atomic *x*-, *y*-, *z*-coordinates, *B*-factors, and occupancies) the data need to be supplemented by additional information. This extra information is applied by way of geometrical restraints. These are target values for geometrical properties such as bond lengths and bond angles and are typically obtained from crystallographic studies of small molecules. The refinement process aims to prevent the bond lengths and angles in the model from drifting too far from these target values, which is achieved by applying additional terms to the function being minimized of the form:

$$\sum_{k=1}^{Distances} w_k (d_{k0} - d_k)^2,$$

where d_k and d_{k0} are the actual and target distance, and w_k is the weight applied to each restraint.

If the structure is refined using full-matrix least-squares refinement, a by-product of this method is that the *s. u.*s of the refined parameters, such as the atomic coordinates and *B*-factors, can be obtained. However, their calculation involves inverting a matrix whose size depends on the number of parameters being refined. The larger the structure, the more atomic coordinates and *B*-factors, the larger the matrix. As matrix inversion is an order n^3 process, it has tended to be unfeasible for molecules the size of proteins and nucleic acids; these have several thousand parameters and consequently a matrix whose elements number several millions or tens of millions, which is why *s. u.*s have been routinely published for small-molecule crystal structures, but not for structures of biological macromolecules. It is purely a matter of size.

Recently, however, as faster workstations with larger memories have become available, the situation has started to change, and calculation of atomic errors has become more practicable. Indeed, *s. u.*s are now frequently calculated for small proteins using SHELX, the refinement package originally developed for small molecules, but sadly, the *s.u.* data are still not commonly deposited in the PDB file. So this makes us none the wiser about the precision with which any given atom's location has been determined.

So what is to be done? What information is there on the reliability of an X-ray crystal structure?

What should one look for? First of all, there are several parameters relating to the overall quality of the structure commonly cited in the literature that can be found in the header records of the PDB file itself, as described in Global Parameters for X-ray Structures.

Global Parameters for X-Ray Structures

Resolution

The resolution at which a structure is determined provides a measure of the amount of detail that can be discerned in the computed electron density map. The reflections at larger scattering angles, θ, in the diffraction pattern correspond to higher resolution information coming as they do from crystal planes with a smaller interplanar spacing. The high-angle reflections tend to be of a lower intensity and more difficult to measure and, the greater the disorder in the crystal, the more of these high-angle reflections will be lost. Resolution is related to how many of these high-angle reflections can be observed, although the value actually quoted can vary from crystallographer to crystallographer as there is no clear definition of how it should be calculated. The higher resolution shells will tend to be less complete and some crystallographers will quote the highest resolution shell giving a 100%

complete data set, whereas others may simply cite the resolution corresponding to the highest angle of scatter observed.

The higher the resolution the greater the level of detail, and hence the greater the accuracy of the final model. The resolution attainable for a given crystal depends on how well ordered the crystal is—that

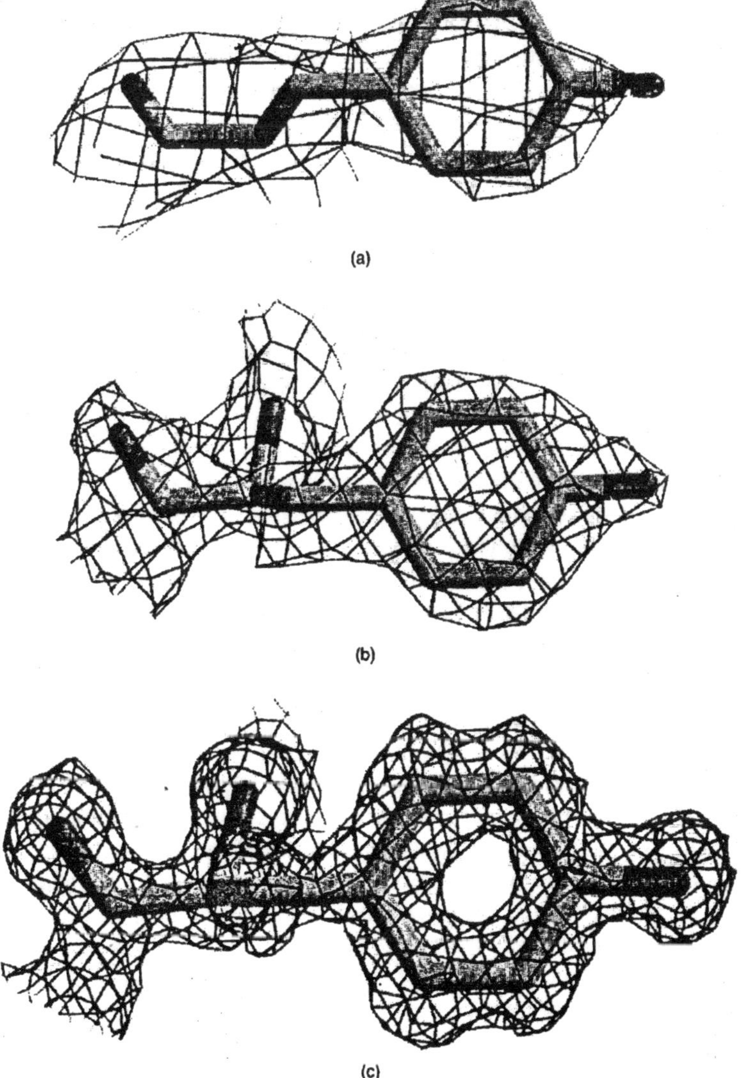

Fig. 7.7. The effect of resolution on the quality of the electron density.

is, how close the unit cells throughout the crystal are to being identical copies of one another. A simple rule of thumb is that the larger the molecule the lower will be the resolution of data collected.

In general, side chains are difficult to make out at very low resolution (4Å or lower), and the best that can be obtained is the overall shape of the molecule and the general locations of the regions of regular secondary structure. Models at such low resolution are clearly of no use for investigating side-chain conformations or interactions! At 3Å resolution, the path of a protein's chain can be traced through the density and at 2Å the side chains can be confidently fitted.

The most precise structures are the atomic resolution ones (from around 1.2Å resolution up to around 0.9Å). Here the electron density is so clear that many of the hydrogen atoms become visible, and alternate occupancies become more easily distinguishable. These structures require fewer geometrical constraints during refinement and hence give a better indication of the true geometry of protein structures.

The resolution of structures in the PDB varies from atomic resolution structures to very low resolution structures at around 4.0Å, with a definite peak at around 2.0Å. The lowest quoted resolution as of November 2001 was 30.0Å for PDB entry 1qgc—the structure of the capsid protein of the foot-and-mouth virus, complexed with antibody, and solved by a combination of cryoelectron microscopy and X-ray crystallography. Resolution is probably the clearest measure of the likely quality of the given model. However, bear in mind that, because there is no single definition of resolution, it tends not to be used consistently and its value can be overstated.

R-factor

The *R*-factor is a measure of the difference between the structure factors calculated from the model and those obtained from the experimental data. Higher values correspond to poorer agreement with the data, whereas lower values correspond to better agreement. Typically, for protein and nucleic acid structures, values quoted for the *R*-factor tend to be around 0.20 (or equivalently, 20%). Values in the range 0.40 to 0.60 can be obtained from a totally random structure, so structures with such values are unreliable and probably would never be published. Indeed, 0.20 seems to be something of a magical figure and many structures are deemed finished once the refinement process has taken the *R*-factor to this mystical value.

As a reliability measure, however, the *R*-factor is itself somewhat unreliable. It is quite easily susceptible to manipulation, either deliberate or unwary, during the refinement process, and so models with major errors can still have reasonable-looking *R*-factors. For example, one of the early incorrect structures, cited by Brändén and Jones (1990), was that of ferredoxin I, an electron transport protein. The fully refined structure was deposited in 1981 as PDB code 2fd1, with a quoted resolution of 2.0Å and an *R*-factor of 0.262. Due to the incorrect assignment of the crystal space group during the analysis of the X-ray diffraction data, this structure turned out to be completely wrong. The replacement structure, reanalyzed by the original authors and having the correct fold, was deposited as PDB entry 3df1 in 1988. Its resolution was given as 2.7Å and its *R*-factor as 0.35. On the face of it, therefore, mere comparison of the resolution and *R*-factor parameters would lead one to believe the first of the two structures to be the more reliable! The reason that an *R*-factor as low as 0.262 was achieved for a totally incorrect structure was that the coordinates included 344 water molecules, many extending far out from the protein molecule itself. This is a large number of waters for a protein containing only 107 residues. A rule of thumb suggested by Brändén and Jones (1990) is that, for high-resolution structures, one water molecule for each residue is reasonable, and waters should only be added to the structure if they make plausible hydrogen bonds.

Incidentally, the version of ferredoxin that was 3df1 was itself twice superseded, first by entry 4df1 in mid-1988 and then by entry 5df1 in 1993. The last of these had a quoted resolution of 1.9Å and *R*-factor of 0.215.

The ferredoxin example is one of overfitting; that is, having too many parameters for the experimental data available. It is always possible to fit a model, however wrong, to the data if there is an excess of parameters over observations.

R_{Free}

A more reliable measure is Brunger's free *R*-factor, or R_{free}. This is less susceptible to manipulation during refinement. It is calculated in the same way as the standard *R*-factor and again measures the agreement between the structure factors as calculated from the model and as obtained from the experimental data. It differs in that its calculation uses only a small fraction of the experimental data, typically 5–10%, and, crucially, this fraction is excluded from the structure refinement procedure. The test set, as it is called, thus provides an independent measure of the goodness of fit of the model to the data while the refinement proceeds on the remaining data, the working set.

Unless there are correlations between the data in the test set and those in the working set, the refinement process should not be able to influence the R_{free} measure.

The value of R_{free} will tend to be larger than the R-factor, although it is not clear what a good value might be. Brunger has suggested that any value above 0.40 should be treated with caution. There were approximately 20 structures in this category in the PDB, as of November 2001. Not surprisingly, most are fairly low-resolution structures (3.0–4.0Å).

Average positional error

Even though atomic coordinate *s.u.* s are not commonly given, it is quite usual for an estimate of the *average* positional error of a structure's coordinates to be cited. There are two principal methods for estimating the average positional errors: the Luzzati plot and the σA plot.

The Luzzati plot is obtained by partitioning the reflections from the diffraction pattern into bins according to their value of sin θ, where θ is the reflection's scattering angle, and then calculating the R-factor for each bin. The value calculated for each bin is plotted as a function of sin θ/λ, where λ is the wavelength of radiation used. The resulting plot is compared against the theoretical curves of Luzzati (1952) to obtain an estimate of the average positional error. One problem with this method is that the actual curves do not usually resemble the theoretical ones at all well, and so the error estimate is somewhat crude and often merely provides an upper limit on the error. Better results are obtained if the R_{free} is used instead of the traditional R-factor.

The σ_A plot provides a better estimate still. It involves plotting ln σ_A against (sin $\theta/\lambda)^2$, where σ_A is a complicated function that has to be estimated for each (sin $\theta/\lambda)^2$ bin, as described in Read (1986). The resultant plot should give a straight line whose slope provides an estimate of the average positional error.

Most refinement programs compute both error estimates from the Luzzati and Read methods, so these values are commonly cited in the PDB file. You will find them in the file's header records under the now unfashionable term "*estimated standard deviation*" (or ESD).

Bear in mind that an average *s.u.* is exactly what it says: an average over the whole structure. The *s. u.*s of the atoms in the core of the molecule, which tends to be more ordered, will be lower than the average, while those of the atoms in the more mobile and less well-determined surface—and often more biologically interesting—regions will be higher than the average.

Atomic B-factors

A more direct, albeit merely qualitative, way of determining the precision of a given atom's coordinates is to look at its associated *B*-factor. *B*-factors are closely related to the positional errors of the atoms, although the relationship is not a simple one that can be easily formulated. It is safe to say, however, that atoms in a structure with the largest *B*-values will also be those having the largest positional uncertainty. So if high levels of precision are required in your analysis, leave out the atoms having the highest *B*-factors. As a rule of thumb, atoms with *B*-values in excess of 40.0 are often excluded as being too unreliable. Similarly, if atoms in your region of interest, such as an active site, are all cursed with high *B*-factors then your region of interest is not well determined and you will need to be careful about the conclusions you draw from it.

Rules of Thumb for Selecting X-Ray Crystal Structures

Many analysis in structural bioinformatics require the selection of a dataset of 3D structures on which to perform one's analysis. A commonly used rule of thumb for selecting reliable structures for such analyses, where reasonably accurate models are required, is to choose those models that have a quoted resolution of 2.0Å or better, and an *R*-factor of 0.20 or lower. These criteria will give structures that are likely to be reasonably reliable down to the conformations of the side chains and local atom–atom interactions. One example that uses such a dataset is the Atlas of Protein Side-Chain Interactions, which depicts how amino acid sidechains pack against one another within the known protein structures.

Of course, the selection criteria depend on the type of analysis required. For some analyses only atomic resolution structures (i.e., 1.2Å or better) will do, as in the accurate derivation geometrical properties of proteins—for example, side-chain torsional conformers and their standard deviations, or fine details of the peptide geometry in proteins that can reveal subtle information about their local electronic features. For other types of analysis, structures solved down to 3Å may be good enough, as in any comparison of protein folds. One interesting example is that of the lactose operon repressor. Three structures of this protein were solved to 4.8Å resolution, giving accurate position for only the protein's C^{α} atoms. However, because the three structures were of the protein on its own, of the protein complexed with its inducer, and of the protein complexed with DNA, the global differences between the three structures showed how the protein's conformation changed between its induced and repressed states. Thus even very low resolution structures were able to help explain how this particular protein achieves its biological function.

Often the above rule of thumb (resolution ≤2.0Å, and *R*-factor ≤0.20) is supplemented by a check on the year when the structure was determined. Structures are more likely to be less accurate the older they are simply because experimental techniques have improved markedly since the early pioneering days of the 1960s and 1970s. Indeed, many of the early structures have been replaced by more recent and accurate determinations.

Error Estimates in NMR Spectroscopy

The theory of NMR spectroscopy does not provide a means of obtaining *s.u.* s for atomic coordinates directly from the experimental data, so estimates of a given structure's accuracy and precision have to be obtained by more indirect means.

Global parameters for NMR structures

As mentioned above, a number of models can be derived that are compatible with the NMR experimental data. It is difficult to distinguish whether this multiplicity of models reflects real motion within the molecules or simply results from insufficient experimentally derived restraints. Generally, the agreement of NMR models with the NMR data is measured by the agreement between the distance and angular restraints applied during refinement of the models and the corresponding distances and angles in the final models. Large numbers of severe violations would indicate a serious problem of data interpretation and model building.

However, the errors associated with the original experimental data are sufficiently large that it is almost always possible to generate models that do not violate the restraints, or do so only slightly. Consequently, it is not possible to distinguish a merely adequate model from an excellent one by looking for restraint violations alone.

Traditionally, the quality of a structure solved by NMR has also been measured by the root-mean-squared deviation (rmsd) across the ensemble of solutions. Regions with high rmsd values are those that are less well defined by the data. In principle, such rmsd measures could provide a good indicator of uncertainty in the atomic coordinates; however, the values obtained are rather dependent on the procedure used to generate and select models for deposition. An experimentalist choosing the best few structures for deposition from a much larger draft ensemble can result in very misleading statistics for the PDB entry. For example, the best few structures may, in effect, be the same solution with minor variations—so the rmsd values will be small. Structures further down the original list may provide alternative solutions, which are slightly less consistent with the data, but that are radically different. The sizes of ensembles deposited in the PDB range from 1 to 85 models.

The number of experimentally derived restraints per residue can give an indication of how effectively the NMR data define the structure in a manner analogous to the resolution of X-ray structures. Indeed, the number of restraints per residue correlates with the stereochemical quality of the structures to an extent, but some restraints may be completely redundant and no consistent method of counting is used by depositors. None of these measures gives a true indication of the accuracy of the models, that is, how well they represent the true structure, and few are reported in the PDB file.

In recent years, NMR equivalents of the crystallographic *R*-factor have been introduced. One method involves the use of dipolar couplings. These provide long-range structural restraints that are independent of other NMR observables such as the NOEs, chemical shifts, and couplings constants that result from close spatial proximity of atoms. Because the expected dipolar couplings can be computed for a given model, they provide a means of comparing observed with expected, and obtaining an *R*-factor that is a measure of the difference between the two. What is more, it is also possible to obtain a cross-validated *R*-factor, equivalent to the crystallographic R_{free}, wherein a subset of dipolar couplings are removed prior to the start of structure refinement and used only for computing the *R*-factor. This gives an unbiased measure of the quality of the fit to the experimental data. However, in the case of NMR, one cannot use a single test set of data; one has to perform a complete cross-validation. The reason for this is that, whereas in crystallography each reflection contains information about the whole molecule, in NMR each dipolar coupling does not. So a complete cross-validation is required, which means that a number of calculations have to be performed, each using a different selection of test sets and working data sets; the test set, which usually comprises 10% of the whole data set, being selected at random each time.

Another technique for calculating an NMR *R*-factor uses the NOEs and involves back-calculation of the NMR intensities from the models obtained and comparison with those observed in the experiment. This technique is implemented in the program RFAC, which calculates not only an overall *R*-factor for the entire structure, but also local *R*-factors, including residue-by-residue *R*-factors and individual *R*-factors for different groups of NOEs (e.g., medium-range NOEs, long-range NOEs, interresidue NOEs, etc.).

An additional back-calculation method for checking structure quality is to calculate the expected frequencies (positions) of spectral peaks from the structure and compare them to those observed. This comparison has the advantage that the frequencies are not usually a target of the structure refinement procedure. However, the measures described here are not yet generally included in the deposited PDB files.

Rules of thumb for selecting NMR structures

Historically, the rule of thumb for selecting NMR structures for inclusion in structural analyses has been the simple one of excluding them altogether! This early prejudice stems from the fact that they were viewed as being of generally lower quality than X-ray structures, there was no easy way of selecting them with a consistent rule as that used for selecting X- ray structures, and they represented only a minority of the PDB anyway. However, nowadays NMR structures provide much valuable information about protein and DNA structures not available from X-ray studies. Indeed, although only about one in eight PDB structures come from NMR experiments, in data sets of representative structures around one in four are NMR structures. This stems from the fact that many unique and important proteins can only be solved by NMR. Nevertheless, it is still not possible to differentiate between reliable and unreliable NMR structures from the information given in the PDB files. There is no standardized information provided that is akin to the resolution, *R*-factor, and estimated *s.u.*s routinely quoted for X-ray crystal structures. The only way to get an idea of the quality of the structure is to read the paper describing it and judge from the statistics provided there or, more ambitiously, to carry out your own analysis of either the stereochemistry of the structure or the agreement between restraints and structures in those cases where the experimental data has been deposited along with the structure.

Errors in Deposited Structures

Serious Errors

There have been a number of serious errors in X-ray and NMR structures documented in the literature. Many of the erroneous models have been retracted by their original authors, or replaced by improved versions. Structures are often re-refined, or solved with better data, and the models in the PDB are replaced by the improved versions. The models that are replaced do not completely disappear, though. There is a growing graveyard of obsolete structures—some very, very incorrect, others merely slightly mistaken—available at the Archive of Obsolete PDB Entries. This Web site provides a graphic history of each structure, some of which have gone through several incarnations.

Of all errors, the most serious are those where the model is, essentially, completely wrong; for example, the trace of the protein chain follows the wrong path through the electron density and the resultant model has the wrong fold completely. There is practically no similarity between the correct and incorrect models.

The next most serious errors are where all, or most, of the secondary structural elements have been correctly traced, but the chain connectivity between them is wrong. Here the erroneous model has most of the correct secondary structure elements, and has them arranged in the correct architecture. However, the protein sequence has been incorrectly traced through them (in one case going the wrong way down a β-strand). Thus most of the protein's residues are in the wrong place in the 3D structure. Such errors arise because the loop regions that connect the secondary structure elements tend to be more flexible, and more disordered, so their electron density tends to be more poorly defined and difficult to interpret correctly. This situation was particularly true in the case as the *primary sequence* of the protein was unknown at the time the structure was being solved and had to be guessed from the limited clues in the electron density map.

Less serious are frame-shift errors, although they can often result in a significant part of the model being incorrect. These errors occur where a residue is fitted into the electron density that belongs to the next residue. The frame shift persists until a compensating error is made when two residues are fitted into the density belonging to a single residue. These mistakes often occur at turns in the structure, and almost exclusively at very low resolution (3Å or lower).

The least serious model-building errors involve the fitting of incorrect main-chain or side-chain conformations into the density. Of course, even such errors, depending on where they occur, can have an effect of the biological interpretation of what the structure does and how it does it.

Typical Errors

Typically, the models deposited in the PDB will be essentially correct. The remaining errors will be the random errors associated with any experimental measurement. As mentioned above for X-ray structures, the average *s.u.*s—estimated on the basis of the Luzzati and σ_A plots—can provide an idea of the magnitude of these errors. The values range from around 0.01Å to 1.27Å. Note that the latter value approaches the length of some covalent bonds! The median of the quoted *s.u.*s corresponds to estimated average coordinate errors of around 0.28Å. It has to be remembered that these values are estimates, and apply as an average over the whole model.

Stereochemical Parameters

An alternative way of assessing a structure's quality, which complements the types of checks described so far, is to examine its geometry, stereochemistry, and other structural properties. A number of tests can be applied to a protein or nucleic acid structure that compare it against what is known about these molecules. This knowledge comes from systematic analyses of the existing structures in the PDB. In other words, the vast body of structures that have been solved to date provides a knowledge base of what is normal for proteins and nucleic acids.

The advantage of such tests of normality is that they do not require access to the original experimental data. Although it is possible to obtain the experimental data for many PDB entries—structure factors in the case of X-ray structures, and distance restraints for NMR ones—these entries are still the minority, and deposition of these data is still at the discretion of the depositors. Furthermore, to make use of the data requires appropriate software packages and expert know-how. The stereochemical tests, however, require no experimental data. So any structure, whether experimentally determined, or the result of homology modeling, molecular dynamics, threading, or blind guesswork can be checked. The software is freely available and easy to use and interpret. What is more, many of the results of such checks made on existing structures are readily available on the Web, as will be mentioned below.

Most of the tests described here apply exclusively to protein structures. Similar tests have been developed for DNA and for small molecules (hetero atom groups) that may be bound to protein or DNA. These will be mentioned later. The stereochemical tests include bond lengths, bond angles, torsion angles, hydrogen bond energies, and so on.

Before describing the checks, one crucial point needs to be stressed at the start. The majority of the checks compare a given structure's properties against what is the norm. Yet this norm has been derived from existing structures and could be the result of biases introduced by different refinement practices. Furthermore, outliers, such as an excessively long bond length or an unusual torsion angle, should not be construed as errors. They may be genuine—for example, as a result of strain in the conformation, say, at the active site. The only way of verifying whether oddities are errors or merely oddities is by referring back to the original experimental data. Indeed, the experimenters who solved the structure may already have done this, found the apparent oddity to be correct and commented to that effect in the literature.

Having said that, if a single structure exhibits a large number of outliers and oddities, then it probably does have problems and can safely be excluded from any analyses.

Proteins

Ramachandran plot

Perhaps the best-known, and certainly the most powerful, check for the stereochemical quality of a protein structure is the Ramachandran plot. This plot is of the ψ main-chain torsion angle versus the ϕ main-chain torsion angle for every amino acid residue in the protein (except the two terminal residues, because the N-terminal residue has no ϕ and the C-terminus has no ψ). In the resulting scatter plot, the points tend to cluster in certain favorable regions, and tend to be excluded from certain disallowed regions due to steric hindrance of the side-chain atoms. Glycine and proline, which have no side chains as such, have slightly different distributions on the plot, although they too have regions from which they are excluded.

The favorable regions correspond to the regular secondary structures: right-handed helices, extended conformation (as found in β-strands), and left-handed helices. Even residues in loops tend to lie within these favored regions. The residues show a tight clustering in the most favored regions with few or none in the disallowed regions. The regions themselves have been determined from an analysis of torsion angles in existing structures in the PDB. It comes from a structure that shall remain nameless. Here the majority of the residues lie in the disallowed regions, and it can be confidently concluded that the model has serious problems.

One caveat concerns proteins containing D-amino acids rather than the more common L-amino acids. These residues have the opposite chirality so their ϕ–ψ values will be negative with respect to their L-amino cousins. The Ramachandran plot for D-amino acids is the same

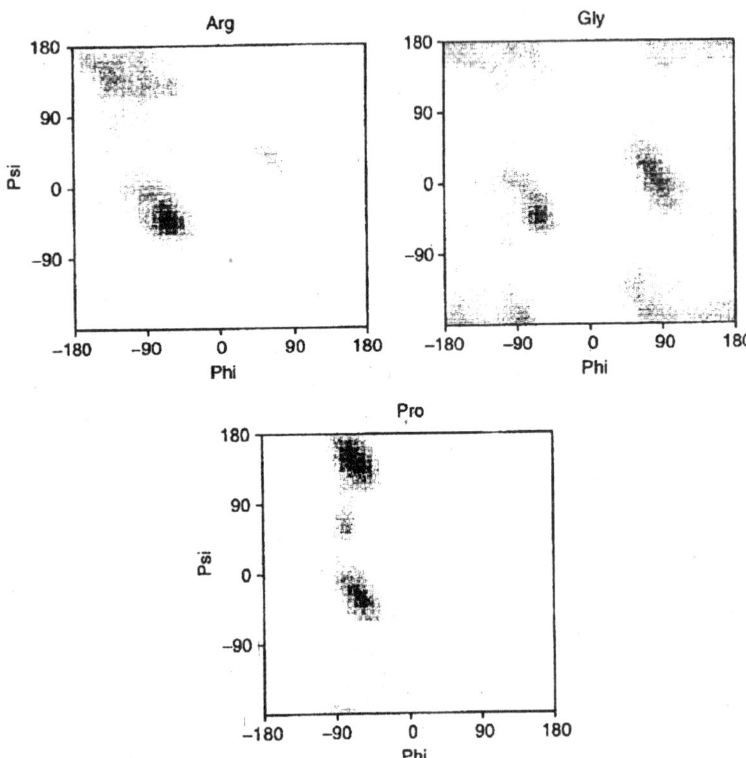

Fig. 7.8. Differences in Ramachandran plots for (a) arginine, representing a fairly standard amino acid residue, (b) glycine which due to its lack of a side chain, (c) proline which due to its restraints on the movement of the main chain has a restrictec range of ϕ values.

as for L-amino acids, but with every point reflected through the origin. Thus, proteins such as gramicidin A (e.g., PDB code 1grm) that have many D-amino acids, give Ramachandran plots that look particularly troubling but that may be perfectly correct.

The tightness of clustering tends to be a function of resolution, with atomic resolution structures exhibiting very tight clustering. At lower resolution, as the data quality declines and the model of the protein structure becomes less accurate, so the points on the Ramachandran plot tend to disperse and more of them are likely to be found in the disallowed regions.

One feature that makes the Ramachandran plot such a powerful indicator of protein structure quality is that it is difficult to fool (unless one does so intentionally by, say, restraining ϕ–ψ values during structure refinement as is sometimes done for NMR structures). This reliability was demonstrated by Gerard Kleywegt in Uppsala who once attempted to deliberately trace a protein chain *backwards* through its electron density to see whether it would refine and give the sorts of quality indicators that could fool people into believing it to be a reasonable model. Of the parameters that he tried to fool, the two that seemed least gullible were the R_{free} factor mentioned above and the Ramachandran plot. The latter looked most unhealthy, with several residues in disallowed regions and no significant clustering in the most highly favored regions.

A simple measure of quality that can be derived from the plot is the percentage of residues in the most favorable or core regions. (Glycines and prolines are excluded from this percentage because of their unique distributions of available ϕ-ψ combinations). Using the core regions defined by the PROCHECK program, one generally finds that atomic resolution structures have well over 90% of their residues in these most favorable regions. For lower and lower resolution structures this percentage drops, with structures solved to 3.0–4.0Å tending to have a core percentage around 70%. NMR structures also show increasing core percentage with increasing experimental information. However, NMR structures can have relatively good side-chain positions even with a poor core percentage as NMR data restrain side chains more strongly than the backbone because of the large number of side-chain protons.

Side-chain torsion angles

Protein side chains tend to have preferred conformations, known as rotamers, about their rotatable bonds, again as a result of steric hindrance. The rotamers are defined in terms of the side-chain torsion angles χ_1, χ_2, χ_3, and so on. The first of these, χ_1, is defined as the torsion angle about N—C^{α} —C^{β}—A^{γ}, where A^{γ} is the next atom along the side chain (for example, in lysine the A^{γ} atom is C^{γ}). The next, χ_2, is defined as C^{α} —C^{β}—A^{γ} —A^{δ}, and so on. The χ_1 and χ_2 distributions are both trimodal with the preferred torsion angle values being termed gauche-minus (+60$^{a\%}$), trans (+180$^{a\%}$), and gauche-plus (“60$^{a\%}$). A plot of χ_2 against χ_1 for each residue has 3 × 3 preferred combinations, although the strength of each depends very much on the residue type.

Like the Ramachandran plot, a plot of the χ_1-χ_2 torsion angles can indicate problems with a protein model as these, like the ϕ and υ torsion angles, tend not to be restrained during refinement. .What is more, these torsion angles tend to cluster more tightly toward their ideal rotameric values as resolution improves. For example, the standard deviation of the χ_1 torsion angles about their ideal position tends to be around 8° for atomic resolution structures and can go as high as 25° for structures solved at 3.0Å. Similarly, the corresponding standard deviations for the χ_2 torsion angles tend to be 10° and 30°, respectively.

Bad contacts

Another good check for structures to be wary of is the count of bad and unfavorable atom–atom contacts that they possess. Too many and the model may be a poor one. The simplest checks are those which merely count bad contacts, that is, those where the distance between any pair of nonbonded atoms is smaller than the sum of their van der Waals radii. Furthermore, the atoms checked should not merely be those involved in intraprotein contacts within the given protein structure; for X-ray crystal structures it is also necessary to consider atoms from molecules related by crystallographic and noncrystallographic symmetry.

More sophisticated checks consider each atom's environment and determine how happy that atom is likely to be in that environment. For example, the ERRAT program counts the numbers of nonbonded contacts, within a cutoff distance of 3.5Å, between different pairs of atom types. The atoms are classified as carbon (C), nitrogen (N), and oxygen/sulfur (O), so there are six distinct interaction types: CC, CN, CO, NN, NO, OO. If the frequencies of these interaction types differ significantly from the norms (as obtained from well-refined high-resolution structures) the protein model may be somewhat suspect. A similar analysis can be used to locate local problem regions by using a nine-residue sliding window and obtaining the interaction frequencies at each window position.

One level up in sophistication is the DACA method, which is implemented in the WHAT IF program. DACA stands for Directional Atomic Contact Analysis and compares the 3D environment surrounding each residue fragment in the protein with normal environments computed from a high- quality data set of protein structures. There are 80 different fragment types, including main-chain fragments as well as side-chain fragments. The environment of each fragment is essentially the count of different nonbonded atoms in each 1Å × 1Å × 1Å cell of a 16Å × 16Å × 16Å cube surrounding the fragment. A similar approach is that of the ANOLEA program, which calculates a nonlocal energy for atom–atom contacts based on an atomic mean force potential.

Other parameters

Other parameters that can be used to validate protein structures include counts of unsatisfied hydrogen bond donors and hydrogen-bonding energies as is done in the WHATCHECK program mentioned below.

C-alpha only structures

As of November 2001, there were around 200 structures in the PDB (out of over 16,000) that contain one or more protein chains for which only the C^{α} coordinates have been deposited. The deposition of C^{α}-only coordinate sets is usually done where the data quality has been too poor to resolve more of the structure. It was common in the early days of protein crystallography for only C^{α}s to be deposited; nowadays it is still quite common for only C^{α}s to be deposited for very large structures, such as the recently determined structure of the ribosome at 5.5Å. The standard validation checks are of no use for such models, lacking as they are in so much of their substance. However, there is an equivalent to the Ramachandran plot for these structures. The parameters plotted are the C^{α}—C^{α}—C^{α}—C^{α} torsion angle as a function of the C^{α}—C^{α}—C^{α} angle for every residue in the protein. As with the Ramachandran plot, there are regions of this plot that tend to be highly populated, and others that appear forbidden. So a structure with many outliers in the forbidden zones should be treated with caution. The checks are incorporated in the program MOLEMAN2 which can be run over the Web.

Nucleic Acids

Finding validation tools for DNA and RNA is trickier than for proteins. The PDB's validation tool, ADIT (AutoDep Input Tool), incorporates a program called NuCheck for validating the geometry of DNA and RNA. Binary versions of the ADIT package can be downloaded for use on SGI and Linux machines. A program specifically developed for checking the geometry of RNA structures, but that can also be used for DNA structures, is MC-Annotate. It computes a number of peculiarity factors, based on various metrics including torsion angles and root-mean-square deviations from standard conformations, that can highlight irregular regions in the structure that may be in error or merely under strain.

Hetero Groups

The geometry of hetero compounds, as deposited in structures in the PDB, tends to be of widely varying quality. The HETZE program is one of the few validation methods that checks various geometrical parameters of the hetero compounds associated with PDB structures.

These parameters include bond lengths, torsion angles, and some virtual torsion angles, the information principally coming from the small-molecule structures in the CSD.

Software for Quality Checks

A large number of programs are freely available that can perform the sorts of quality checks described above on proteins, nucleic acids, and hetero compounds. Below are listed the most commonly used programs not requiring any specialist knowledge or additional specialist software.

PROCHECK

PROCHECK computes a number of stereo- chemical parameters for a given protein model and outputs the results in easy-to-understand colored plots in PostScript format. Significant deviation in the parameters from the standards that have been derived from a database of well-refined high- resolution proteins are highlighted as being unusual. The plots include: Ramachandran plots, both for the protein as a whole and for each type of amino acid; χ_1-χ_2 plots for each amino acid type; main-chain bond lengths and bond angles; secondary structure plot; deviations from planarity of planar side chains; and so on.

WHATCHECK and WHAT IF

The WHATCHECK program is a subset of Gert Vriend's WHAT IF package. It contains an enormous number of checks and produces a long and very detailed output of discrepancies of the given protein structure from the norms. The DACA method, mentioned above, for analyzing nonbonded contacts, is incorporated into the original WHAT IF program.

PROVE

PROVE compares atomic volumes against a set of precalculated standard values. Volumes are calculated using Voronoi polyhedra to define the space that each atom occupies by placing dividing planes between it and its neighbors.

SQUID

The SQUID program displays two-dimensional and three-dimensional data derived from protein structures using many graph types. It can also be used for validation via ready-to-use scripts.

ERRAT

The ERRAT program has already been described. It analyzes nonbonded atom contacts in protein structures in terms of CC, CN, CO, and so forth contacts.

Quality Information on the Web

Rather than having to install and run one of the above packages, it is possible to obtain much of the information they provide from the Web. Several sites provide precomputed quality criteria for all existing structures in the PDB. Other sites allow you upload your own PDB file, via your Web browser, and will run their validation programs on it and provide you with the results of their checks.

PDBsum—PROCHECK Summaries

The first site that provides precomputed quality criteria is the PDBsum Web site at http://www.biochem.ucl.ac.uk/bsm/pdbsum. This Web site specializes in structural analyses and pictorial representations of all PDB structures. Each structure containing one or more protein chains has a PROCHECK and a WHATCHECK button. The former gives a Ramachandran plot for all protein chains in the structure, together with summary statistics calculated by the PROCHECK program. These results can provide a quick guide to the likely quality of the structure, in addition to the structure's resolution, *R*-factor and, where available, R_{free}. The WHATCHECK button links to the PDBREPORT for the structure, described below.

Occasionally the model of a protein structure is so bad that one can tell immediately from merely looking at the secondary structure plot on the PDBsum page. Most proteins have around 50–60% of their residues in regions of regular secondary structure, that is, in α-helices

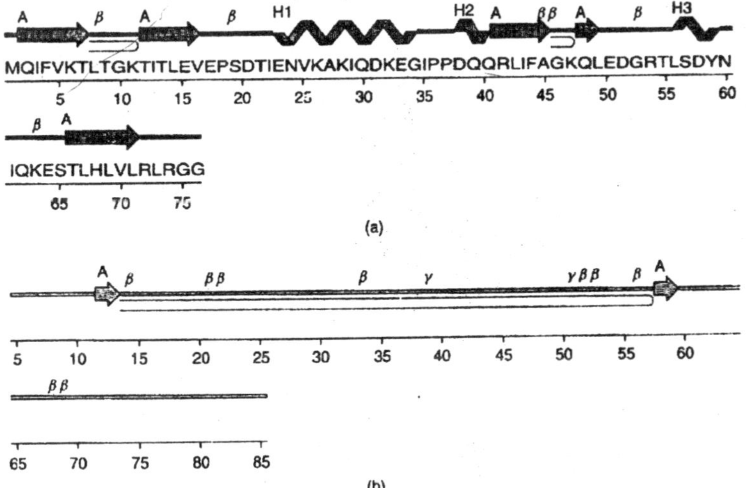

Fig. 7.9. Schematic diagram of two protein models in the PDB.

and β-strands. However, if a model is really poor, the main-chain oxygen and nitrogen atoms responsible for the hydrogen-bonding that maintains the regular secondary structures can lie beyond normal hydrogen-bonding distances; so the algorithms that assign secondary structure may fail to detect some of the α-helices and β-strands that the correct protein structure contains.

PDBREPORT—WHATCHECK Results

The WHATCHECK button on the PDBsum page leads to the WHAT IF Check report on the given protein's coordinates. This report is a detailed listing of the numerous analyses that have been precomputed using the WHATCHECK program. These analyses include space group and symmetry checks, geometrical checks on bond lengths, bond angles, torsion angles, proline puckers, bad contacts, planarity checks, checks on hydrogen-bonds, and more, including an overall summary report intended for users of the model.

PDB's Geometry Analyses

The PDB Web site also has geometrical analyses on each entry, consisting of tables of average, minimum, and maximum values for the protein's bond lengths, bond angles, and dihedral angles. Unusual values are highlighted. It is also possible to view a backbone representation of the structure in RasMol, colored according to the Fold Deviation Score—the redder the coloring the more unusual the residue's conformational parameters.

Validation Servers on the Web

In addition to the sites mentioned above, there are a number of validation servers on the Web that allow you to submit a PDB file for analysis. They are mostly for protein structures and most use programs that are freely available for in-house use. However, the servers can often be easier and more convenient to use, and of course save you having to download and install the programs, particularly the Biotech Validation server that runs the three most commonly used validation programs: PROCHECK, PROVE, and WHATCHECK.

The main aim of this section is to impress on you that the macromolecular structures that form the very foundation of structural bioinformatics are not all of the same quality and can undermine that foundation if not carefully selected. All structures are just models devised to satisfy data obtained experimentally. As such, they will contain errors, both systematic and random. Some structures have been found to be seriously incorrect, that is, they are inaccurate models of the molecules they represent and in many cases have been replaced by more accurate models.

Most structures are reasonably accurate but inevitably contain random errors, as is symptomatic of any experimental measurement. The quality of structures as a whole has improved over the past few years and this trend is expected to continue. However, determining which is a good structure and which is not is still not straightforward. Even traditional measures, such as the resolution and *R*-factor for X-ray structures, and number of restraints for NMR structures, do not always separate the good from the bad. Very often, other quality measures need to be taken into account when selecting a good data set.

Structure Comparison and Alignment

Structure comparison refers to the analysis of two or more structures looking for similarities in their three-dimensional (3D) structures. Alignment refers to establishing equivalences between amino acid residues based on the 3D structure of two or more protein folds. All commonly used methods do a reasonably good job of recognizing the more obvious instances of similar 3D folds, but, as we shall see, structure alignments are much more variable. Most algorithms and resulting Web resources provide protein structure comparison and alignments at the level of domain or complete polypeptide chain.

It is important to clear up immediately any confusion between structure comparison and alignment versus structure superposition. These terms are sometimes used interchangeably in the literature; here, we make a stricter delineation. Structure superposition assumes that you already know of at least some residues that match between protein structures A and B. Typically, these C-alpha positions then become anchor points and the task becomes one of using a minimization technique or analytical procedure to find a transformation that minimizes the distance between aligned residues. The best solution is then the one that produces the lowest root mean square deviation (rmsd) between A and B. In other words, the alignment is already assumed and all that is required is tweaking to bring the two structures into register. Clearly, this is a much easier problem (and there is an exact solution to it) than having no a priori knowledge of what amino acids are equivalent. The latter is known as the structure alignment problem.

Finding proteins that exhibit 3D similarity and then finding the best alignment through gap insertions is much more difficult, and forms the basis of this chapter. Clearly structure superposition could play a role once the alignment is complete and the relationship between residues in the two structures has been established.

WHY IS 3D STRUCTURE COMPARISON AND ALIGNMENT IMPORTANT?

The first question to ask is why is structure comparison and alignment important? This question has been answered several times in other chapters in this book; for example:

1. Structure classification methods use structure alignments to help in the assignment of fold classes and can be used subsequently in establishing libraries of templates for use in proteome annotation.
2. Structure alignments of a protein of known fold and function against a protein of unknown function can provide insight into the function of the unknown. Structure alignments are of particular importance in an era of genomically driven structure determination where no a priori knowledge of the biological function may exist.
3. Structure prediction methods require that the predicted structure be evaluated against a variety of template structures.
4. Structural alignments reveal distant sequence relationships not available from sequence alignments alone and can be used in protein engineering and protein modeling.

As a general point, it is worth highlighting again here, even though it is introduced elsewhere in this book, that structure alignments provide us with information not available from current sequence alignment methods. The reason for this is the result of nature's ability to reduce complexity to manageable levels yet still maintain incredible diversity and adaptability. If you consider that an average protein consists of 300 amino acids, then there are 20^{300} possible proteins—more than the number of atoms in the universe. Nature has selected a very small subset of these—as few as 30,000 in humans for the functioning of a complex organism. Still greater reduction exists in three dimensions—all proteins from all species are believed to be represented by somewhere between 1000 and 5000 protein folds. This remarkable reduction was first noted in the globins when only a small number of structures existed. It was later manifest in the hssp curve, which was recently updated by Rost (1999).

Here, we take a slightly different look at the relationship using data from our own laboratory. Each point on the graph represents one of 1000 randomly selected polypeptide chains taken from the Protein Data Bank (PDB) that show a measure of structure similarity measured

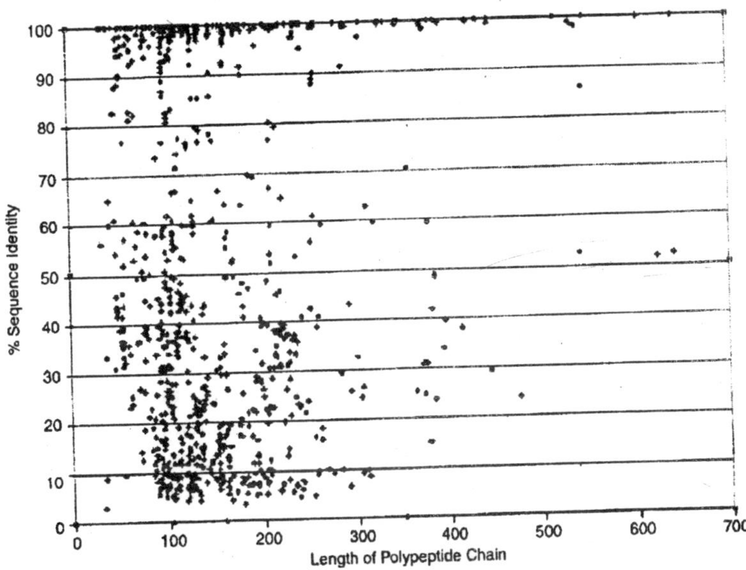

Fig. 7.10. Structure similarity versus sequence similarity.

by CE with a z-score greater than 4.5, indicating structure similarity at the level of the protein superfamily. Plotted on the x-axis is the length of the polypeptide chain and on the y-axis the resulting sequence identity. Thus, while there are a number of matches with very high sequence identity (90% or greater) indicating post-translational modifications in the PDB, there are also many chains with low sequence identity. The region between 30% and 20% sequence identity indicates a twilight zone where this relationship may be detectable by sequence methods alone. However, below 20% sequence identity is the so-called midnight zone where only structure comparison reveals the relationship between these two proteins at the level of the fold. Rost (1999) using an even larger data set showed that this relationship between sequence and structure, while having a length dependency, can be represented as a Gaussian distribution with a peak around the 9% sequence identity level. Certainly the relationship between these proteins, let alone an accurate sequence alignment, could not be achievable by sequence methods alone. Thus, structure alignments provide valuable insights not achievable from sequence alone.

As discussed elsewhere, it is dangerous to consider these findings as absolute; they most certainly are not. The relationship between primary protein sequence, 3D structure, and biological function is more complex and still being interpreted. As George Bernard Shaw once said, "the golden rule is that there are no golden rules." For example, there are cases of structures that contain large regions of high sequence similarity yet share no structure similarity. The viral capsid protein (1PIV: 1) shares an 80-residue stretch with glycosyltransferase (1HMP:A) where there is greater than 40% sequence identity, yet the structures within those regions are completely different (mostly beta versus mostly alpha, respectively).

General Approach to Structure Comparison and Alignment

Structure comparison and alignment is an NP-hard problem that is solved heuristically by all methods. Although different heuristics employed by different methods tend to recognize similar folds, they will not provide exactly the same structure alignments. In fact two structure comparison methods may produce alignments that differ in every position. We consider the impact of an NP problem and the fact that even if it were a tractable problem, it may provide the best analytical but not biological answer, after discussing the details of the most popular structure comparison and alignment methods.

There is a significant body of literature on methods of pairwise protein structure comparison and alignment. Orengo (1994) provided an overview of the field until 1994. Gibrat et al., (1996) highlighted some surprising results from structure alignment using different methods and Lemmen and Lengauer (2000) summarized protein structure alignment in the context of the general problem of structure alignment and superposition in drug design. In reading these reviews of various methods you will see that each methodology that has been tried can be boiled down to three or possibly four steps depending on how you count (steps 2 and 3 can be considered a single step):

1. Represent the proteins A and B (polypeptide chains, domains, or other amino acid fragments) in some coordinate independent space so that they can be readily compared
2. Compare A and B
3. Optimize the alignment between A and B
4. Measure the statistical significance of the alignment against some random set of structure comparisons

This methodology applies to pairwise structure comparison and alignment. Multiple structure alignment involves a somewhat different methodology and is considered separately. Given this general approach there are two classes of problem that the defined algorithms try to solve. The first is to optimize the alignment between any given pair of proteins; the second is given a new target structure to determine, in some rank order, which structures in the PDB are most like the target.

Pragmatically, even if a comparison between two proteins takes 30 seconds on a typical processor, today this still represents a significant computation with approximately 18,000 proteins in the PDB representing approximately 30,000 chains, some multidomain, resulting in over 428 years of computation. This compute bottleneck is solved in two ways by several of the resources. First, as structures are released by the PDB each week, they are added to an all-by-all comparison database and so the computations are performed incrementally. Notwithstanding, even with 50 to 100 new structures appearing each week this is still a significant computation.

Second, the known relationship between sequence and structure is employed to reduce the number of computations to be performed. A rough estimate is that only 1 in 10 new structures represents a new fold. This ratio is dependant on the method used and how one defines a fold, but it serves to provide a rough estimate. To establish that ratio of 1:10, at least 5 of the 10 similarities can be inferred from sequence alone without the need for structure comparison. Hence, this reduces the number of targets that need to be checked. Similarly, PDB structures can be grouped into a set of structural representatives so that the target is only compared against a subset of the complete PDB. To better understand the concept of a representative consider the CE algorithm, described subsequently, which uses the following criteria to define a single representative for a number of polypeptide chains being represented:

1. The rmsd between two chains is less than 2Å
2. The length difference between two chains is less than 10%
3. The number of gap positions in alignment between two chains is less than 20% of aligned residue positions
4. At least two-thirds of the residue positions in the represented chain are aligned with the representing chain

Other methods apply similar rules, but at the domain level. This reductionism somewhat reduces the accuracy of the comparison but provides the necessary gains in speed.

The discussion here considers steps 1–4 (steps 2 and 3 are discussed together) for several of the most popular methods of protein structure comparison and alignment. By popular we refer to methods that have been cited a significant number of times and for which Web resources exist for the reader to immediately access these methods, and importantly, these resources are kept current. Web resources are of two types. First, there are those that allow you to compare two proteins using the specific method, and, second, there are those that provide a database of precalculated comparisons against all or a subset of the PDB. These two types are not mutually exclusive—you may be able to look up an existing database of comparisons and alignments and submit your own structure for comparison and alignment to the same resource. It is beyond the scope of this chapter to deal with each method in detail. The intent is to give the reader a sense of the similarities and differences in approaches that have been employed.

Protein Structure Representation

As stated, the first step is to suitably represent the two proteins to be compared. Certainly the methods presented here are not exhaustive for this step. For example, geometric hashing taken from computer vision is a technique applied by Nussinov and Wolfson (1991) and later refined by Fischer et al., (1994), but not part of any of the methodologies discussed.

Also at issue is what is being compared. Since domains are the functional units of currency for proteins, it makes sense to compare domains. Problems do arise, there is not always agreement on what constitutes a domain. However, a method such as CE, which focuses on polypeptide chains, can miss recognizing a domain if the domain is part of a long chain for which no similarity is found in other parts. Such a result reduces the significance of the match of those two domains.

DALI

Pragmatically, in terms of community availability and use (something these authors rate highly) while significant work had been done beforehand, notably from Taylor and Orengo (1989), protein structure comparison was popularized by the 1993 paper of Holm and Sander (1993a). This paper describes the use of distance matrices as embodied in the DALI method. Based on these numbers it is fitting to start with DALI.

DALI uses distance matrices to represent each structure to be compared. This idea was not new and dates back to the work of Phillips (1970). Each structure is represented as a two-dimensional (2D) array of distances between all C-alpha atoms. This representation has the advantage of placing all structures in a simplified common frame of reference. Conceptually, the problem is then straightforward, as if one is imagining each structure's contact map transparently overlaid. Overlap along the diagonal then represents similar backbone conformations (secondary structure) and off-diagonal similarity tertiary structure similarity. Moving one sheet of paper horizontally or vertically relative to the other to achieve overlap represents gap insertion into one or other of the structures. In a later version, a quick look-up was introduced that uses the alignment of secondary structure elements (SSEs) and is not dissimilar to VAST.

CE

The combinatorial extension (CE) algorithm also uses a distance approach to structure comparison, but at the level of octameric fragments; that is, a comparison of C-alpha distance matrices is made for every combination of eight residues in each protein chain. Each octameric fragment that aligns within the two structures being compared, based on a heuristic measure, is referred to as an aligned fragment pair (AFP).

COMPARER

COMPARER uses the comparison of residues' properties, segments (of residues), and relations between residues, and relations between segments. Examples of residue properties are identity and local conformation. Examples of segment properties are secondary structure type and orientation relative to the center of gravity. Examples of residues relationships are hydrogen bonds and hydrophobic clusters. Examples of segments relationships are distances to one or more nearest neighbors and the relative orientation of two or more segments.

SARF2

SARF2 operates at the level of SSEs represented by the C-alpha atoms of each residue. First, SSEs are detected by comparison with typical helix and strand templates to within 0.4Å and 0.8Å, respectively. Second, compatible pairs of SSEs and larger assembles of SSEs are constructed and analyzed.

SSAP

SSAP uses the comparison of intraprotein C^{β}-C^{β} vectors (calculated using dummy C_{β} atoms in the case of glycine) to provide directionality.

VAST

The Vector Alignment Search Tool (VAST), as the name suggests, represents structures as a set of SSEs (vectors) whose type, directionality, and connectivity infer the topology of the structure.

Comparison Algorithm and Optimization

DALI

The distance matrices are collapsed into regions of overlap (submatrices) of fixed size, which are then stitched together if there is overlap between adjacent fragments. A solution to the branch and bound algorithm for finding the overlap and optimal superposition is neatly described in Holm and Sander (1996), a later paper than that describing the original method.

CE

CE uses three thresholds in the alignment-building process. The first threshold detects AFPs. The second threshold evaluates the suitability of a next candidate AFP relative to the current alignment (a single AFP in the beginning). The third threshold evaluates all alignments to find those that are optimal. The second and third thresholds define whether new AFPs are added to the alignment. Alignment extension is sought in a narrow area of the search space limited to a single gap of no longer than 30 residues in either of the two proteins being compared.

This restriction permits computational tractability, but may miss nontopological alignments and also those with insertions of more than 30 residues. These thresholds are empirical, being based on observed comparison of intraresidue distances in structures known to align. If one or more significant alignments are found (up to 30 top-scoring alignments are retained), then further optimization is performed using dynamic programming and interprotein distances calculated based on superposition. The last step is repeated iteratively until the optimal alignment is found.

COMPARER

Fourteen properties and relationships are selected. For properties the dynamic programming algorithm is used to find the optimal alignment. For relationships the dynamic programming is not applicable since there is a dependence of scores for a given relationship on the assignment of other relationships. For this reason the so-called combinatorial simulated annealing technique is used for optimization.

SARF2

Pairs of SSEs are evaluated for the angle between them, the shortest distance between their axes, the closest point on the axes, and the minimum and maximum distances from each SSE to their medium line. Then, searching for the largest ensembles of the mutually compatible pairs of SSEs is performed using an algorithm from graph theory used to solve the *maximum clique problem*. Further refinement and extension of the alignment by incorporation of additional residues is then performed.

SSAP

SSAP finds the optimal structure alignment by applying dynamic programming to the matrix of scores S_{ik} for every pair of positions i and k from two proteins A and B, respectively. S_{ik} is obtained from comparison of vectors between C^{β} atoms at pairs of positions i and j to C_{β} atoms in *selected matching positions*.

The selected matching positions are in turn defined by applying dynamic programming to the matrix of differences of C^{β}-C^{β} vectors from positions i and k to all other protein positions. Since dynamic programming is applied at two levels the whole procedure is called *double dynamic programming*.

VAST

Given the SSEs, alternative alignments are examined using a Gibbs sampling algorithm, beginning with a seed SSE-pair alignment. The optimal alignment is defined as that which is most surprising relative to the background distribution of alpha-carbon superposition residuals obtained by chance.

Statistical Analysis of Results

DALI

The similarity score is derived from an all-against-all comparison of 225 representative structures with less than 30% sequence identity. The DALI score is then expressed as the number of standard deviations (z-score) from the average score derived from the database background distribution.

CE

Two distributions of rmsds and gaps are built and numerically tabulated for the 25% nonredundant set. The final z-score is calculated by combining z-scores from two tabulated distributions under the assumption of their normality.

COMPARER

Two scores E and A are introduced to measure residues' equivalence and gap penalties, respectively. Both scores are calculated based on scores seen in two unrelated proteins and two random sequences relationships.

SARF2

The similarity score is calculated as a function of rmsd and the number of matched C-alpha atoms. The significance of a particular comparison is evaluated by comparison of the score with the score distribution built from the comparison of the protein leghemoglobin with a set of 426 nonredundant structures.

SSAP

So-called raw SSAP scores derived from the comparison are calibrated against known comparisons in the Classification, Architecture, Topology, Homology (CATH) database. Thus, raw SSAP scores above 70–80 are indicative of topology level similarity if 60% of the residues of the larger protein are included in the alignment.

VAST

The significance of a VAST match is determined in a manner similar to its sequence counterpart, BLAST. VAST calculates a p-value for the best substructure superposition as the probability that this score would be seen by chance in drawing SSE pairs at random, multiplied by the number of alternative substructure alignments possible, given the SSEs in the protein pair under consideration.

How Well Are We Doing?

Using the preceding methods and others, our ability to recognize common folds, not anticipated from sequence alone, has led to interesting biological findings. A sample is presented in Sample Results from Structure Comparison and Alignment below. However, in using structure alignment as a tool in structural bioinformatics the answer is not so straightforward.

Gaining new insights into such areas as structure prediction and biological function derived from remote homologs through structure alignment and not decipherable from sequence alignment alone is compelling. However, as Godzik (1996) has shown with respect to various structure alignment methods, we have shown in aligning the catalytic subunit of the protein kinases, and Jones has shown with respect to the comparison of structure classification methods, significant differences in structure alignments exist.

Differences in structure alignments are not surprising given the heuristics that each methodology applies to make an NP-hard problem computationally tractable. A simple view of these differences is as follows: Consider a spectrum that at one end maximizes the geometric relationship between two proteins and at the other provides the maximum amount of biological significance in the alignment. Depending on the task at hand, you may wish to be at one end of the spectrum or the other, or in the middle. Favoring the geometric end of the spectrum will likely lead to a better rmsd but more fragments, that is, a larger number of gaps and a loss of biological relevance.

An example of this would be the breaking of hydrogen bonds in a beta sheet to better fit fragments of those sheets to each other in the two proteins being compared. Favoring the biological end of the spectrum will likely lead to a higher rmsd. Which methods favor which ends of the spectrum? An easy question, with a not so easy answer, since the answer is dependant on the parameters used when computing with each method and the particular proteins under study.

An important consideration when using any structural alignment method is to consider the nature of the problem you are trying to solve and to experiment with a variety of methods.

To illustrate the importance of the above statement, consider some results from our own work in comparing expert hand alignments of protein kinases against those produced by CE where the goal is to achieve the best functionally relevant alignment. The gold standard was the hand alignment of 18 protein kinases all with sequence identity below 40%—a substantial challenge to any sequence alignment method. In comparison to the gold standard, CE failed to make the best biological alignment in every case, partly because of significant spatial movements of secondary structural elements (as well as loops) relative to each other in the various structures.

However, these observations do not necessarily suggest that methods treating secondary structures as rigid bodies will yield superior results. In considering the alignment between cAMP-dependant protein kinase (1CDK:A) and an actin-fragmin kinase (1CJA:A) where the sequence identity is 13% and there is significant structure diversity, CE correctly inserted a gap in a beta strand to better preserve the orientation of side chains.

Conversely, in the same structure pair, an aspartic acid residue, which is functionally critical as the catalytic base in the phosphotransfer reaction, is misaligned. This misalignment occurs as a result of the residue being adjacent to a loop region that presents a difficult challenge to CE. The reader is referred to the paper for further examples. What is clear is that better scoring functions are needed that can better incorporate what is known about the structures and function(s) of the respective molecules being fed into the alignment.

Sample Results from Structure Comparison and Alignment

As methods of structure comparison and alignment were published, they bought forth many previously unobserved structure comparisons, that were later captured in such resources as SCOP, CATH, and the DALI domain dictionary. We consider one such example here. Later, as structure comparison and alignment methods matured they became a standard experimental method integral to attempting to understand biological function. Here, we consider a second example taken from our own work where structure comparison was used in conjunction with other supporting evidence to provide a putative biological function subject to further experimental analysis. Together these two examples illustrate the importance of structure comparison and alignment to biology.

Holm and Sander (1993b) showed that the membrane insertion domain of the bacterial toxin colicin A has the same topology of fold as the globins and phycocyanins with six helices sequentially aligned. Both in terms of sequence and function there is no relationship and the original authors missed this structural similarity. The implication is that this similar fold represents structural convergence to a stable three-on-three helical sandwich.

An example from our own work illustrates the coming together of information from different methodologies, including structure comparison and alignment, to define a putative function. In this case the methodologies employed hidden Markov models (HMMs), site information from Prosite, and structure comparison found with CE. It was found that the alpha–beta hydrolase fold family that includes acetylcholinesterases contain putative Ca^{2+} binding sites, which in some family members may be critical for heterologous cell associations. From a structure or sequence comparison perspective alone the evidence was not definitive. Structure alignments between acetylcholinesterases and classic calcium binding EF-hands, as found, for example, in calmodulin, are apparent but tentative—3Å to 4Å rmsd over approximately 80 residues with 10–20 residue gaps.

Nevertheless, the Prosite signature for calcium binding is present and HMMs reveal potential calcium-binding sites for cholinesterases as well as neuroligins and gliotactins. It is postulated that with extracellular Ca^{2+} concentrations higher in the extracellular matrix than within the cell, binding associations could be weaker. While the outcome of this analysis is still in question, it does point to the need for experiments, for example mutagenensis at the site of calcium binding and monitoring of subsequent enzyme activity. Such synergy between in silico and in vivo and in vitro experiment is the future of structural bioinformatics.

Multiple Structure Alignment

Our discussions thus far have involved only pairwise structure comparison and alignment, or at best, alignment of multiple structures to a single representative in a pairwise fashion—progressive structure alignment. True multiple structure alignment attempts to align all structures simultaneously to provide the best consensus alignment between all structures, which may not be the best alignment between any pair of structures. In principle, if it were accurate enough, multiple structure alignment could enhance the impact that profiles and HMMs have had from a purely sequence perspective by providing multiple alignments with weak yet definitive sequence relationships. A few approaches to multiple structure comparison and alignment have been undertaken. Here, we outline one method to illustrate the principle of multiple structure alignment and compare the results to accurate hand alignments.

Our approach uses Monte Carlo optimization of an existing set of pairwise alignments derived using CE, and hence is referred to as MC-CE. From the starting alignment—a set of structures all aligned to a master structure—a set of moves were designed to address different alignment situations in a similar manner to that previously used by Mirny and Shakhnovich (1998) for sequence–structure alignment. Moves are then applied in a random manner to a constrained search space to seek the optimal alignment. Each step is tested against a scoring function and accepted with some level of probability. This procedure proceeds until convergence, that is, no random steps improve the optimal alignment as based on a distance score for each block of aligned residues across the multiple structures. Running MC-CE against 66 protein families produced a 12% increase in the number of aligned columns and a 22% decrease in total alignment length when compared to pairwise alignments. When compared to the hand alignments of the HOMSTRAD database and our own hand alignments for the protein kinase family there was widespread agreement, particularly in the more rigid C-terminal lobe of the protein kinase catalytic subunit where substrate recognition and binding take place. The N-terminal lobe is more flexible and challenging. Consider one specific example to illustrate the issues, namely, the two beta strands of the N-terminal lobe that contain the glycine-rich loop. This region of the kinase domain is flexible and often in different conformations depending on the state of the enzyme. However, it contains the well-conserved GxGxxG motif, which is important for the binding of ATP in the active site. With the exception of one structure (1CJA:A) it should be aligned without gaps to properly align this motif. Standard CE alignment splits off some of the sequence leading up to one strand and unnecessarily separates off a row of conserved glycines in the loop between strands. The MC-CE alignment compresses the sequence leading up to strand one, and closes the gap, which causes the glycine displacement in CE. However, MC-CE does not correct the misaligned glycine residues seen in some structures in the original CE pairwise alignments.

This example illustrates that multiple structure alignment techniques are in their infancy and that resources such as HOMSTRAD, with their human curation of specific protein families, are very valuable.

Mapping Protein Fold Space

Knowing that there are a finite number of folds and that at this time experimentally we have a reasonable percentage of those folds, estimated at 25–50%, and that structural genomics is aimed at giving us as many of the rest as possible, it is not surprising that we have attempted to establish "maps" of protein fold space, which we will continue to fill in. Structural Classification of Proteins (SCOP) and CATH are perhaps the best examples of those maps when other data, for example, sequence homology are included, but what about considering nothing but structure? The ultimate and almost certainly unanswerable question is, can we establish a structure- based phylogenetic tree that evolved from a single common ancestor—the original protein fold? Problems arise immediately since there are different views on what constitutes a protein fold. The vertical axis is a count of the number of aligned residues broken into different cells of sequence identity and rmsd across the complete PDB as determine using the CE algorithm introduced previously. The rectangular slab of cells with greater than 20% sequence identity and in the 1–2Å rmsd range illustrate why comparative modeling works in defining a protein structure from sequence. What is surprising is the large number of residues in the 3–6Å rmsd range with very low (<20%) sequence identity. We have previously analyzed this information in one interpretation of protein fold space. Subsequently, we used the alignments of these regions that average 80 resides in length and are contained within domains to determine what specific conserved residue properties exist in these common substructures. Although some of these substructures are well known, others have not been described explicitly in the past and yet occur frequently. Perhaps of most importance, these substructures overlap—part of one will be found in another, indicating the continuity of protein fold space. Thus, one answer to the question of what exactly constitutes a protein fold is a discrete and discernibly reused element of protein structure as determined systematically through comparative structure analysis. The history of characterizing protein folds has not been so systematic, but very valuable nevertheless. References such as Greek key motif, the jelly roll, the beta propeller, and so forth are well accepted yet were not defined systematically. Resources such as SCOP and CATH have undertaken a more systematic approach. Yet, if protein fold space is a continuum, then many systematic definitions are possible. The only point to make then to the student of protein structure space is to carefully define how you have systematically chopped up that space and be aware of the characterizations of the space that have been made before you. In short there are great opportunities for original thinking in defining and using protein fold space, particularly as that space gets more populated in the coming years.

For those who have already dabbled in the study of protein structure space, the challenge is to construct a map of the space at the right resolution to solve the problem at hand. For example, we continue to work with relatively high resolution maps, where many would call the discreet units substructures rather than folds, to address problems in protein engineering. Others work with lower resolution maps—folds and larger structures.

Lower resolution maps often address the evolution of protein folds and must distinguish between convergent and divergent evolution. For protein engineering this distinction does not matter; it is the characteristics of the fold, notably its stability under mutagenesis that is of concern. Distinguishing between convergent and divergent evolution of protein folds must take into account sequence and/or functional considerations. We consider two characterizations of fold space that consider evolutionary constraints.

Partslist has many features, but includes mapping of folds to fully sequenced genomes by homology modeling, thereby providing a distribution of folds within a diverse set of species. Partslist reveals that both families, superfamilies and folds follow a power-law distribution. In short, considering folds, a small number of folds are used many times and there is a steep fall-off where many folds are used only occasionally. Simpler organisms share these highly utilized folds.

Conversely, Holm and Sander (1996) have examined all of existing fold space—which they admit, and we emphasize here, is biased by the current contents of the PDB—and defined what they refer to as attractors in shape space. They hypothesize that attractors represent both dominant folding pathways and evolutionary sinks as a result of physical constraints. The result is five dominant attractors containing 40% of all known folds in 16 different fold classes.

The charting of protein fold space has strong analogies to the history of global maps. At present we would seem to be working at the level of Ptolemy's map. This map was designed specifically for navigation, yet missed many of the vital elements. Structural genomics may well be the protein structure equivalent of Columbus's voyage opening a new world to be explored and utilized.

The impetus for improved protein structure comparison, alignment, and characterization will be defined quite simply by quantity—the rate of increase in the number of experimentally determined new folds and the number of structures that conform to each fold. Hand comparisons and alignments will require automation while still retaining the level of quality provided by experts today, since no expert will be able to keep up with manual comparisons and alignments. Today we are in a mode where tools assist in the process with final decisions about comparisons and alignments made by human experts. This situation will not scale into the future. Quantity implies not just numbers but variety, complexity, and singularity. We can anticipate the structures of more membrane proteins, which because of issues of solubility are today under-represented in the PDB. Complexity implies more structures of biological assemblies determined by conventional means. Finally, singularity will be defined by the structures of a large number of single-domain proteins determined by structural genomics from which protein–protein interactions will be ascertained and that will form pieces of a puzzle to be fitted to an outline defined by lower resolution techniques such as cryoelectron microscopy.

Several years ago one of us estimated that by the end of 2005 the PDB would contain 35,000 structures, almost double the number present at the time of writing. While this number remains optimistic, structural genomics is moving from the engineering to the production phase so the number will increase rapidly.

In short the impetus to study and contribute to the field of structure comparison and alignment is already here. The result will be more structure alignment databases, either generic or comprised of specific folds, families, and superfamilies that will be tools for further research. Consider some examples of that research, some of which is already ongoing:

1. Faster and more accurate protein fitting to electron density maps using consensus alignments.
2. Improved functional characterization of proteins derived from structural genomics.
3. Use of structure-based profiles and hidden Markov models (HMMs) to enhance sequence-only methods to find more distant sequence homologs.
4. Better understanding of the relationship between sequence and fold in specific protein families.
5. Better protein structure prediction through better fold libraries used in both homology modeling and fold recognition.

8

Drug Discovery

Modern pharmaceutical discovery has benefited from both the rigor of scientific discovery and the acceleration of technological advancements. The pharmaceutical industry had its origins at the beginning of the twentieth century. Scientific advancement over the past 100 years has seen the discovery of DNA, the understanding of proteins as specific molecular entities, and the harnessing of X-rays to understand proteins at the atomic level. Structural bioinformatics now is poised to do its part to accelerate the drug discovery process. This chapter follows the historic development of the current paradigm for pharmaceutical drug discovery, and highlights how structural bioinformatics is influencing this process.

Historical Development of Drug Discovery

The current dominant paradigm in pharmaceutical drug discovery seeks to find a particular small molecule inhibitor to bind to a specific receptor, a macromolecular target. Our ability to pursue this paradigm rests on scientific and technological achievements in the twentieth century, particularly with regard to our ability to manipulate organic small molecules on the one hand, and to study the biological targets on the other.

Humanity has, of course, been looking for remedies for its ailments long before there was a drug discovery industry. The use of willow bark as a treatment for pain relief, for example, can be traced back to Hippocrates and earlier. Such use is entirely empiric—a certain recipe gave relief to certain symptoms. Many such folk remedies were known, the progeny of some of which, such as willow bark, have found their place in our modern medicine chests.

The first step toward our modern approach to drug discovery was the suggestion that these remedies generally contained an active ingredient that could be isolated and purified. This idea can be traced back to 1530 to Paracelsus, a Swiss physician. It would be nearly another 300 years, in 1829, however, before the active ingredient in willow bark, salicin, was purified.

The synthesis the year before of urea by Fredrich Wöhler ushered in organic synthesis, which gave chemists the ability to manipulate these small organic compounds. Salicylic acid itself was first synthesized in 1852.

Along with the power to create a specific molecule came the ability to create many closely related compounds. These techniques were first put to profitable use in the dye industry in the mid 1800s, creating for the first time numerous low-cost dye compounds. Using such dyes for histological staining, Paul Ehrlich recognized that related molecules often exhibit related biological effects, a concept referred to today as SAR or structure-activity relationships. This concept was applied to derivatives of salicylic acid to try to discover forms of the drug that were less unpleasant for the patient. This research eventually led to the development of acetylsalicylic acid in 1897 by Felix Hoffmann at Bayer. Bayer named this compound Aspirin: "a" for acetyl, "spir" from Spiraea ulmaria, the meadowsweet plant, and "in," a common suffix for medicines at the time.

In noticing how some compounds more readily stained bacterial cells than human cells, Paul Ehrlich eventually developed another major cornerstone of modern drug discovery, the concept of a *therapeutic index*. All drugs have a minimal dose at which they demonstrate beneficial effects, and a minimal dose at which they demonstrate harmful effects. The therapeutic index is simply the ratio of these two doses. The interplay between trying to make compounds more effective and trying to make them safer is one that continues be at the center of pharmaceutical discovery to this day.

The recognition of activity being associated with specific molecular entities was paralleled (much later) by John Langley's suggestion in 1878 that there must be specific "*receptors*" for such compounds in the host, which bind to these entities. Knowing that there is a host receptor however is not the same as knowing what that receptor is. It would be another 100 years, for example, before John Vane and his colleagues discovered the link between aspirin and prostaglandin synthesis, establishing *cyclooxygenase* (COX) as aspirin's site of action.

This work earned John Vane the 1982 Nobel Prize in Physiology or Medicine. Cyclooxygenase was first given a structural face by Michael Garavito and colleagues in 1994. This structure, and that of the inducible COX-2, has made possible the first forays into structure-based design against this venerated target.

Modern Drug Discovery

Presently, most pharmaceutical drug discovery programs begin with a known macro- molecular target, and seek to identify a suitable small molecule modulator. The advent of the postgenomic era has started to point the way to novel targets, about which more will be said later. Typically, however, the target (usually a protein) has already been identified through biological or genetic investigations to be important in the disease of interest. The approach of modern drug discovery is rational and reductionist with a defined hypothesis of how the chosen mechanism of action could be beneficial against disease.

Following the identification of the target of interest, confidence in the approach is built with a variety of genetic and chemical target validation experiments. The process to discover a lead molecule begins with the development of an assay to look for modulators (either inhibitors, antagonists, or agonists) of the target's activity, followed by a *high-throughput screen* (HTS) of a large number of small molecules, in some cases up to a million or more. In the best cases, this method identifies one or more small molecule "hits" in the micromolar range, that is, having binding constants from 10 micromolar to the low nanomolar range.

Elaboration of the initial small molecule hit through medicinal chemistry is next used to try to improve the potency, ideally lowering the K_i to the low nanomolar range to produce a potent lead molecule. There are an estimated 10^{62} possible small molecules, of which obviously only a tiny fraction can be created and tested. Recently, the techniques of combinatorial chemistry have been developed to rapidly generate hundreds and thousands of derivative compounds from a common scaffold in the hit-to-lead optimization stage. Recent years have seen the successful use of a variety of computational techniques, from quantitative structure activity relationships (QSAR) to computer-aided drug design (CADD) and structure-based drug design.

The process of optimizing the lead molecule into a "candidate" drug is usually the longest and most expensive stage in the drug discovery process (although this is still a fraction of the costs of drug development). Although the candidate is usually an analog of the original lead, it is still considered an art to successfully synthesize and select the exact compound that fulfills all the required properties of potency, absorption, bioavailability, metabolism, and safety. In many ways the lead-to-candidate stage of drug discovery is a multidimensional optimization problem searching within the relatively limited chemical space of analogs of the lead compound.

Following the selection of the candidate molecule, the drug development scientists develop large-scale production methods, and conduct the preclinical animal safety studies. Investigational new drugs (INDs) must pass through a set of three clinical trials: Phase I, a small study on healthy subjects to confirm safety; Phase II, a slightly larger study on a patient population to confirm efficacy; and Phase III, a large study of patients to gather additional information about safety and efficacy. However, even after the medicinal chemist has carefully crafted and balanced the properties of potency, bioavailability, and metabolism, over 90% of the compounds entering clinical trials fail to make it to market, most often due to poor biopharmaceutical properties, toxicity, or lack of efficacy. Due to the attrition of so many potential pharmaceuticals and the rising costs of drug discovery, the average cost to bring each new chemical entity (NCE) to market is estimated to be $770 million.

Impact of Structural Bioinformatics on Drug Discovery

Structural Bioinformatics in a Pharmaceutical Context

Informatics and knowledge-based methods play an important role in the framework of the postgenomic drug discovery paradigm, in support of the traditional roles of screening and medicinal chemistry. Genomics and bioinformatics support genetic methods of target identification and

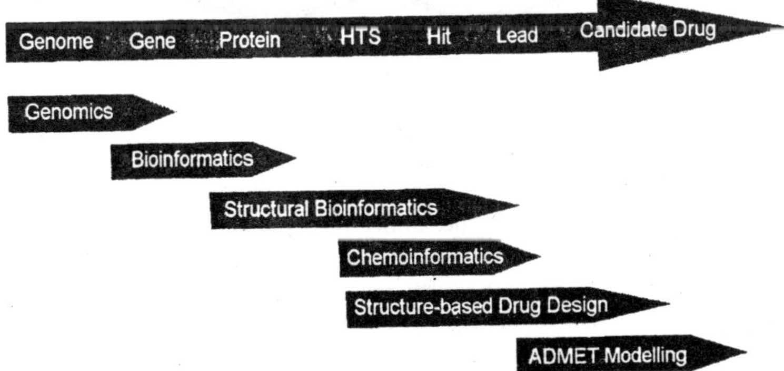

Fig. 8.1. The roles informatics plays in the postgenomic drug discovery.

validation. The ability of chemoinformatics to process the properties of millions of virtual compounds for selection for synthesis and screening is an enabling technology for combinatorial chemistry and HTS. Biological structural information can be usefully exploited from the identification of the target protein all the way to the design of a bioavailable drug via structure-based drug design with suitable drug metabolism properties aided by ADMET (absorption, distribution, metabolism, excretion, and toxicology) modeling.

The techniques of structural bioinformatics are particularly valuable in the area from target identification to lead discovery. Structural bioinformatics group in a pharmaceutical company can serve to link resources and results among bioinformatics, structural biology, and structure-based drug design (SBDD) groups.

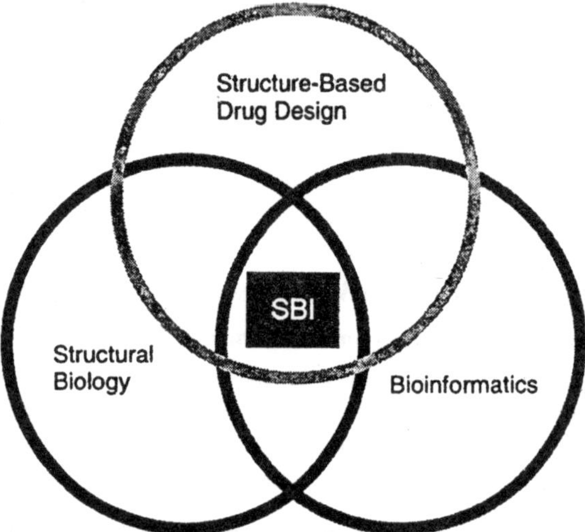

Fig. 8.2. The relationship between structural bioinformatics (SBI) and other disciplines in drug discovery.

Target Assessment

As structural inferences become available at the very earliest stages of a drug discovery program, structural bioinformatics can provide an a priori assessment of the ability of a target to be inhibited by a druglike molecule. Designing compounds with appropriate biopharmaceutical properties that are still able to bind to their targets with an appropriate affinity is the challenge for the medicinal chemist. As detailed below, the challenge for structural bioinformaticians is to determine the magnitude of the medicinal chemists' task.

Assessing target druggability

Not all small molecules can be drugs, and not all proteins can be drug targets. A small molecule must have certain properties, and a protein must contain a binding site that is complementary or compatible with these properties.

Binding sites on proteins usually exist out of functional necessity. Due to hydrophobic forces, the energetically optimal protein would be spherical, with all its hydrophobic residues pointing inward. The majority of successful drugs achieve their activity by competing for a binding site on a protein with an endogenous small molecule. Drugs exploiting allosteric binding sites, with no known natural endogenous ligand, are relatively rare (e.g., the nonnucleoside binding site on HIV-1 reverse transcriptase), and these binding sites are usually not exposed since this is energetically expensive.

Kinases and other ATPases

Examination of the natural ligands of a protein can be valuable in assessing the capacity of a binding site to bind a druglike molecule. The numerous types of ATPases present an interesting example. ATP is a common cofactor for many enzymes. It is recognized by a number of protein folds in a variety of ways. The adenosine portion of ATP (the adenine and ribose rings) has properties one would expect to be able to mimic in a drug. In contrast, it would be difficult to mimic the three charged phosphate groups in a druglike molecule, because charged compounds typically cannot penetrate cell membranes. Thus, when one is considering an ATPase as a potential target it is helpful to determine the way in which the ATP is recognized.

In protein kinases, the adenine ring of ATP fits into a well-defined, relatively hydrophobic pocket, forming a number of important hydrogen bonds. The phosphate groups play a relatively minor role in this recognition. It has proven to be relatively straightforward to generate potent, druglike inhibitors of protein kinases that are competitive for ATP, making use of this attractive binding pocke. In ATPases that rely on coordination of the phosphate group, for example, those containing a so-called Walker A or B motif, inhibition by druglike molecules has proven difficult (except in exceptional cases where, for example, inhibitors bind to two such motifs). In addition, it may also be considerably more difficult to achieve selectivity if the drug predominately exploits main-chain interactions.

In many cases, the exact structure of the ATPase may not be known. Even when the structural motif used to recognize ATP is not known one can gain clues as to the attractiveness of the ATP site by reference to the binding affinities for ATP, ADP, AMP, adenosine, and adenine as gleaned from biochemical studies or from the literature. Differences in dissociation constants in this series may allow one to predict which regions of ATP are important recognition features.

Proteases

The proteases present additional subtleties involved in assessing the tractability of individual molecular targets. The substrates of proteases, that is peptides, represent reasonable chemical leads for a drug discovery process. Crucial to success, however, is the ability to "*depeptidize*" these leads, that is, remove the peptide bonds in order to avoid absorption and metabolic stability issues. For some proteases, this is relatively straightforward. One example is the serine protease thrombin, where many druglike inhibitors, barely resembling peptides, have been developed. In the case of the aspartyl protease renin, it has proven much more challenging to develop nonpeptidic inhibitors.

This difference in druggability can be explained by analyzing the way in which substrate is recognized by these two types of proteases. Serine proteases typically recognize both main-chain and side-chain features in their substrates, forming hydrogen bonds to relatively few main-chain groups on only one side of the scissile bond. Aspartyl proteases are typically tolerant of side-chain substitutions in their substrates and rely on main-chain hydrogen bonds to bind their substrates. In this case, it has proven difficult to retain potency in an inhibitor while reducing the peptidic character of leads. This has also been the case for many viral proteases, where substrate peptides are often weakly bound in shallow surface depressions on the enzyme surface.

Quantitative assessment of target druggability

The foregoing discussion assumes the opportunity to perform an in depth individual analysis of the relevant protein structures. For a quick assessment of a large number of potential targets, a more quantitative approach may be more appropriate.

Such a quantitative approach is already well established for assessing the druglike properties of a small molecule. The rule-of-five is a set of properties to suggest which compounds are likely to show poor absorption or permeation, since such compounds are unlikely to show good oral bioavailability.

As a receptor binding site must be complementary to a drug, it is reasonable to assume that equivalent rules could be developed to describe physicochemical properties of binding sites with the potential to bind rule-of-five compliant inhibitors with a potent binding constant (e.g., K_i < 100 nM). A number of properties complementary to the rule of five can be calculated; for example, the surface area and volume of the pocket, hydrophobic and hydrophilic character, and the curvature and shape of the pocket. Programs such as SURFACE, CAST, ms, and GRASP can be used to calculate these and other parameters. Following the assumption that properties of the drug are complementary to those of the binding site, analysis of the calculated physicochemical properties of the putative drug-binding pocket on the target protein can provide an important guide to the medicinal chemist in predicting the likelihood of discovering a drug against the particular target site.

Table 8.1. The Rule-of-Five

A compound is likely to show poor absorption or permeation if:
1. It has more than five hydrogen bond donors
2. The molecular weight is over 500
3. The Clog P (calculated octanol/water partition coefficient) is over five
4. The sum of nitrogens and oxygens is over 10
5. Weak inhibition (<100 nM) is observed

The logarithmic relationship between the free energy of binding (ΔG) and the binding constant (K_i) means every 10-fold increase in potency is due to a –1.363 kcal/mol change in binding energy

$$\Delta G = -RT\ln(K_i) \quad \ldots(1)$$

Thus a drug with a typical dissociation constant of 10 nM binds with a free energy of –11 kcal/mol and a 1 μM HTS hit binds with –8.4 kcal/mol. The strength of binding is predominately driven by burying of hydrophobic surfaces. The free energy gained from burying hydrophobic surfaces is estimated at around 0.03 kcal/mol/Å^2, with buried polar surfaces giving up about 0.1 kcal/mol/Å^2. A drug with a 10 nM dissociation constant needs to bury 370Å^2 of hydrophobic surface area. Therefore, every 46Å^2 of buried hydrophobic surface (the surface area of a methyl group) buys a 10-fold increase in potency, approximately equivalent to the maximal affinity per nonhydrogen atom defined by Kuntz et al. (1999). Encapsulated cavities are capable of binding low molecular weight compounds with high affinities since they maximize the ratio of the surface area to the volume.

In addition to the predominantly hydrophobic contribution to the binding of many drugs, ionic interactions, such as those found in zinc proteases allow low molecule weight molecules to bind strongly.

Druggable genome

Biological systems contain only four types of macromolecules with which we can interfere using small molecule therapeutic agents: proteins, polysaccharides, lipids, and nucleic acids. Toxicity, specificity, and the inability to obtain potent compounds against the latter three types means that the majority of successful drugs achieve their activity by modifying the activity of a protein by competing for a binding site on a protein with an endogenous small molecule. Thus, there is a limited number of molecular targets for which commercially viable compounds can currently be developed, leading to the concept of the "*druggable genome*." The druggable genome is the subset of the genes in the human genome that express proteins that are capable of binding small druglike (i.e., rule-of-five compatible) molecules.

In a comprehensive review of the accumulated portfolio of targets in the pharmaceutical industry, Drews identified 483 proteins that have been exploited to date. In a critical review of Drews's estimates we have analyzed the sequences of all targets of marketed and investigational drugs or leads that are rule-of-five compatible. Interestingly, only about 120 InterPro domains define all the ligand-binding domains for all proteins for which rule-of-five compliant inhibitors are available.

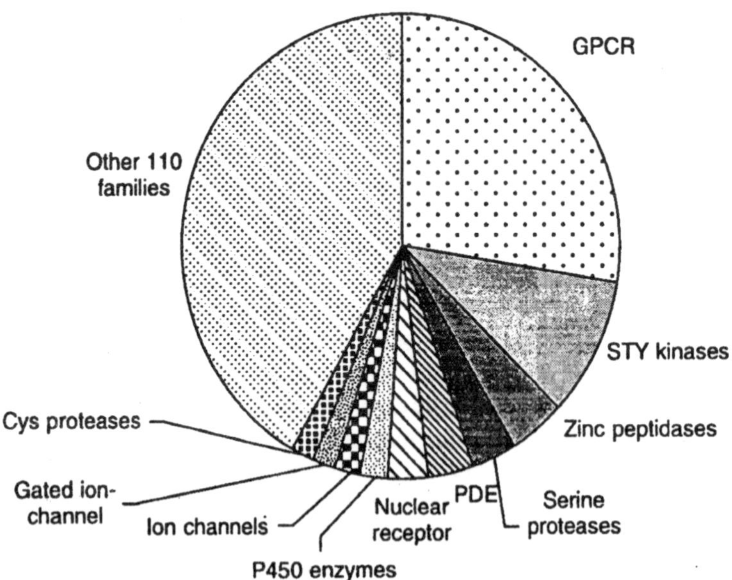

Fig. 8.3. Gene family distribution of the molecular targets of current investigational and marketed drugs.

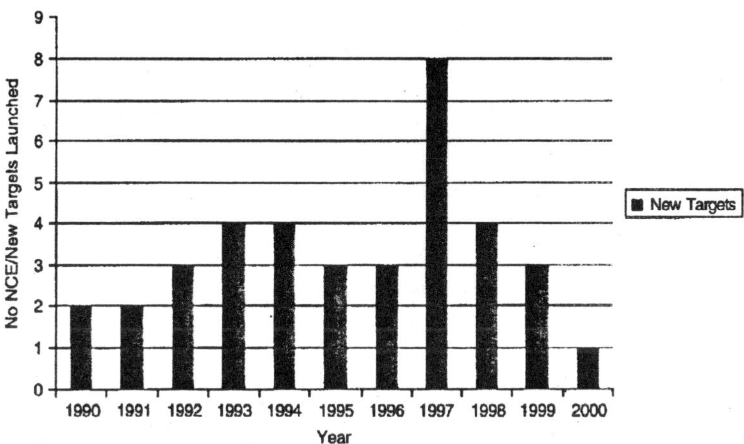

Fig. 8.4. The number of new targets, for which small molecule drugs have been developed and taken to market, launched in the last decade

This distribution of exploited targets actually changes rather slowly. On average, new drugs are launched against only about four novel targets each year.

Existing targets are at the intersection of two necessary attributes: an ability to bind compounds with acceptable properties (druggability), and a link to disease (relevance). The total number of targets that possess both of these attributes can be estimated in a number of ways.

The druggable genome is a subset of the total human genome. The completion of the draft human genome sets the total number of human genes at about 30,000. Estimates of the number of druggable targets in the genome based on the idea of assessing the total number of ligand-binding domains have produced figures in excess of 10,000. Using a more conservative gene family approach, focused on proteins that share greater than 30% sequence identity to a known target in the 120 druggable gene family domains, suggests around 3000 presumed druggable targets in the human genome. Gene families are not equally populated, with genes distributed among a few very large gene families, and many sparsely populated gene families. This distribution suggests that there may be very few large druggable gene families left to discover.

Separately, one can assess how many genes are likely to affect some disease process. Considering about 100 major human diseases, and assuming there are 10 genes directly involved in any given disease process, with another 5 to 10 genes influencing the activity of those genes, yields an estimated 5000–10,000 disease-related genes.

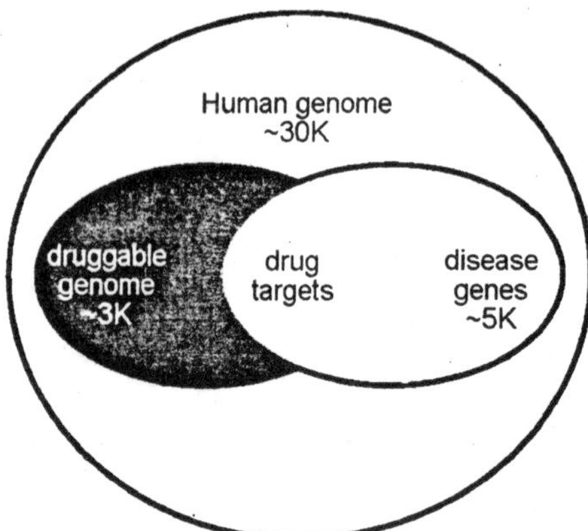

Fig. 8.5. The effective number of exploitable drug targets can be determined by the intersection of the number of genes linked to disease and the druggable of the human genome.

The universe of exploitable small molecule targets for drugs is the intersection between the druggable genome and those genes related to diseases. Structural bioinformatics has a great role to play in identifying all the druggable proteins coded for in genomes of interest.

Target Triage

The availability of the sequences from complete genomes has revealed many more potential targets than could possibly be prosecuted using current experimental technologies. As such it is sometimes necessary to prioritize targets from a large potential subset. Such a subset may arise, for example, from a gene expression study or by analyzing the genomes of disease-causing organisms.

Computational approaches to target triage

A number of these factors can be analyzed computationally. One of the first examples of this is the work of Bruccoleri and others at Bristol-Myers Squibb. In this work, described as congenomics, those genes common to a number of bacterial species but absent from higher organisms were identified as potential targets for antibacterial agents.

This work was taken further at Bayer, where an automated procedure for target prioritization was applied to the recently completed genome of the yeast *S. cerevisiae*, which served as a model for related pathogenic species of fungi. The system developed, CATS (computer-aided target selection), scored targets based on the importance of a gene for the organism, the occurrence of the gene in multiple target species, whether specificity of inhibition could be achieved by reference to sequence similarity with vertebrates, and whether assay development was facile. Of note is that a number of proteins that are the target of antifungal agents scored very highly in this approach, thereby validating this computational approach.

Selectivity

Most potential target have related host proteins whose function must not be affected by a successful therapeutic. This is true, for example, for proteins from the identified gene families, discussed above.

In considering selectivity issues, certainly one must first look at related sequences in the same protein family. However, there is not necessarily a direct correlation between the similarities one would infer from homology and those one would infer from compound action. This discrepancy exists because only a small fraction of the residues in a protein interact directly with any one ligand. Although long-range interactions and conformational changes can play a role in inhibitor binding, in general it is those residues lining a ligand-binding site that are of most importance.

The ATP cleft of the protein kinases serves as a good example of this situation. Although all protein kinases bind ATP in the same region and conformation, only a few of the residues facing the cleft are highly conserved. The remaining residues are subject to random drift, and thus distantly related kinases can actually end up having more similar clefts than more closely related kinases. The tree on the left was constructed using sequence identities calculated across the entire kinase domain for a diverse set of kinase structures. The tree on the right used identities across only the 16 residues lining the ATP-binding

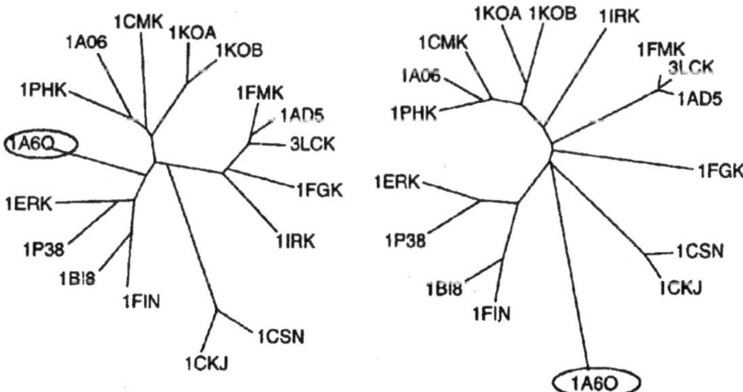

Fig. 8.6. Comparison of phylogenetic trees based either on the entire sequence of the kinase domain or of just the 16 side chains that interact with the ATP.

site. Note for example that the *Zea mays* CK2 kinase, 1A6O, which is grouped with the cyclin-dependent and MAP kinases on the basis of overall sequence identity, is actually seen to be relatively isolated on the basis of its active site alone. This distinction reflects the remarkable ability of this enzyme to hydrolyze either ATP or GTP, due to a number of unique features at the ATP-binding cleft.

This use of noncontiguous regions of sequences to establish relationships was first introduced by Sandberg et al. (1998) and has recently been described as structure-activity relationship homology. In SARAH, protein targets are grouped both by their sequence similarities and by their ability to bind an array of compounds, such as may be determined from pharmaceutical HTS. SARAH builds on earlier ideas around affinity fingerprinting. As Frye points out, protein targets were often grouped according to their response to ligands before we had sequence information at our disposal.

Target Validation

Once a target has been selected for a drug discovery program, it must be experimentally validated. A validated target is one where the premise that modulating the activity of the target has been proven to affect the disease process. A genetic approach to target validation can use a knock-out of the gene of interest, or use RNA antisense technology to inactivate the gene.

The most rigorous way to validate a target is to use a compound, for example, a known small-molecule inhibitor. Clearly at the start of a drug discovery program such compounds may not be available. For some targets the techniques of chemical genetics give us another route. Here, site-directed mutants of the target protein are made in order to make that target sensitive to an existing compound. Essentially this technique uses protein engineering to do what a medicinal chemist would normally be tasked with. Structural bioinformatics clearly plays a central part in this approach, which relies on successfully predicting a match between an existing compound and engineered binding site.

The origins of chemical genetics lie in work by Cohen and others, in which members of the MAP kinase family were mutated in order to make them sensitive to the well-understood inhibitor SB-203580. Others have since used these techniques on other members of the kinase family.

The most attractive aspect of this approach is that the modified gene can be put back into the organism and the activity of compound compared between wild-type and mutant organisms or cell lines. Thus, it allows us to answer questions such as, "If I generated a selective inhibitor of protein X, would it have the desired effect?"

Lead Identification

After a target has been identified and validated, chemical leads (of low or medium potency) must be identified to supply a starting point for the medicinal chemistry efforts. Such leads can come from knowledge of the natural ligands or from *de novo* design, but most often are found through high-throughput screening.

Assay design

In order to design an assay for a particular target, one obviously needs to know what function that target performs and what ligands it recognizes. In the case of novel targets, structural bioinformatics can be used for function and ligand prediction. As described already, a variety of techniques exist to help deduce the function of a protein given knowledge of its structure. Although such methods seldom pinpoint the exact biochemical activity, they can help define what sorts of experiments should be tried.

As discussed earlier, the function of a protein can be dictated by just a handful of residues in a binding pocket. For example, if we were to try to determine the endogenous ligand that might be recognized by a G-protein coupled receptor we may wish to focus on a selection of residues based on mutagenesis data or structure prediction.

Counter-screens can also be used at this point in the drug discovery program as an experimental means to address selectivity concerns. A structural bioinformatics approach can be used to help select those host proteins that are most likely to bind to inhibitors of the primary target. For example, if the primary target were a kinase, hits identified through HTS might be subsequently tested against a panel of related kinases. As described earlier, the proteins most likely to bind the identified inhibitors may not be those most closely related by homology since ligand binding can be dominated by just those residues in the binding cleft. A structural bioinformatics approach, identifying first the obvious and remote homologs and then comparing the presumed ligand-binding residues, can help guide the selection of appropriate counter screen targets.

Using structural similarity to find chemical leads

In general, if two gene products have similar structures, they may bind similar ligands. This relationship enables one to use chemical leads for one protein target against another Even a drug discovery program against a relatively well-understood target can be "jump-started" by recognizing a relationship between the current target and other targets studied in the company. Discovering such a connection can facilitate the transfer of institutional knowledge across the traditional therapeutic area divisions. For example the early work on the antiviral target HIV protease, benefited from the body of knowledge derived from the search for inhibitors of renin, a distantly related aspartyl protease important in cardiovascular disease. Many other examples of such situations can be found among the targets of known drugs, such as angiotensin-converting enzyme, neutral endopeptidase, and thermolysin. In these cases there is no detectable similarity at the sequence level. However all three proteases bind similar substrates and have been shown to bind similar inhibitors. The structural similarity between two of these enzymes has recently been confirmed by the structure determination of neutral endopeptidase.

Using structural similarity to find chemical leads typically causes concern regarding the potential selectivity resultant compounds will have. In practice, however, where the proteins share less than 30% sequence identity the active sites are usually nonidentical and leads can be optimized toward one particular target through computational and medicinal chemistry approaches.

Virtual screening

Due to the high costs of HTS in terms of staffing and compound stock depletion it is highly desirable to attempt to find leads for a protein target computationally. Protein structure can be used in two distinct ways to search databases of compound structures.

The first is to derive a pharmacophore describing the functionally important sites in a ligand-binding site. These sites can be determined by reference to the protein structures, using software such as GRID, X-site, and hotspots. Structures of compounds in chemical libraries (whether existing or not) can then be computationally assessed as to whether they have the ability to adopt a conformation that matches the pharmacophore. The second method is docking and scoring. Here problems are still encountered in correctly docking a small molecule structure to a protein and calculating the binding energy. Current methods can often enrich a set of compounds by identifying a subset of compounds that are more likely to bind. Although showing much promise, the most reliable methods are slow and challenges still exist in being able to distinguish between compounds that do and do not bind to the target protein.

Chemical library design

Chemical libraries, consisting of hundreds to thousands of related compounds, can be generated through combinatorial chemistry by starting with a scaffold or template and decorating it with a variety of functional groups or monomers. A chemical library designed around an early chemical lead can help establish the SAR for that series, pointing the way to more potent compounds. Protein structure can play a great role in the design of such chemical libraries.

Chemical libraries can be designed by reference simply to binding data for small molecule hits. However, if a known or predicted three-dimensional structure for the target is available, or, even better, the structure of a representative compound correctly bound to the protein, such information can be used to design more efficient chemical libraries. In general far more compounds can be designed or enumerated than can actually be produced and tested. Knowledge of the structure of the active site can focus the design and synthesis toward a collection of compounds more likely to bind. In practice, a pharmacophore is usually first defined from the ligand-binding site as is done in virtual screening. Such a pharmacophore can then be applied either to the selection of monomers from which to build the chemical library or to determine which compounds to actually synthesize from an enumerated list of potential compounds.

Lead Optimization

The longest phase in preclinical drug discovery is the process of deriving a high-potency inhibitor from the chemical lead, while optimizing its physicochemical properties to maximize its chances for success as a drug. By bridging and building on resources in bioinformatics, structural biology, and structure-based drug design, structural bioinformatics can accelerate the quest for a potential drug.

Structural biology for structure-based drug design

The most relevant structure-based drug design programs involve repeated cycles of determining the structure of the target in complex with a number of lead compounds and their analogs. One hurdle in establishing a rapid cycle of crystallography and structure-based design is in actually obtaining crystals of the target protein of sufficient quality.

A genetic construct encoding the exact full-length sequence of a protein is not necessarily the ideal one to use in order to obtain material for screening or structural studies. Firstly, the full-length protein may be large and contain protein domains that are not of relevance to the studies being performed. The protein may express poorly, be insoluble, or fail to crystallize, at least in a time frame compatible with the discovery process.

In order to avoid such problems, structural bioinformatics should be used to design suitable constructs at the outset of the project. As described already (domain assignments), proteins are frequently composed of discrete domains. The computational techniques of domain detection can be combined with experimental techniques of domain assignment (for example, limited proteolysis followed by mass spectroscopy analysis). The designed constructs can be evaluated on the basis of their expression levels, solubility, activity, and, of course, crystallizability.

The selection of an appropriate domain for structural studies often begins by aligning the sequence of the target protein with that of a protein of known structure. One can then determine where it may be possible to truncate the protein at the amino- and carboxy-termini. Where structures are not available, secondary structure prediction can be used. Combining secondary structure prediction based on multiple sequence alignments with analysis of sequence conservation can be particularly successful. One can often express a protein from a construct designed to start and finish where sequence conservation is high and at the end of predicted elements of secondary structure

It should be noted that the same process can be used to guide construct design for the production of the target protein for HTS assays. In fact, in the best cases the same construct can be designed for both screening and structural biology.

As described already (crystallography), a second hurdle in structure determination by crystallography is the so-called "*phase problem.*" One common solution to this problem is molecular replacement. In molecular replacement, a model representing some or all of the new protein is rotated and translated into the new unit cell in an attempt to find a solution. Originally molecular replacement was only used for cases of a specific protein in different space groups, or in cases of high sequence identity. Greater computational power, as well as the greater wealth of known folds, has resulted in molecular replacement being successfully applied even in cases where the starting model exhibited only 20% sequence identity or less. The extent to which the core of a protein is distorted with greater sequence divergence presumably will place a limit on when molecular replacement can be used. However, even when phases are determined experimentally (for example, through heavy atoms or selenomethionine techniques), identification of a suitable structural homolog at even 10% sequence identity can greatly accelerate the interpretation of the electron density maps and of the protein structure.

Use of protein surrogates

Structural bioinformatics can play a key role in structure-based drug design approaches even when the structure of the target protein is not available; for example through the generation of a homology model or the use of a surrogate protein. Clearly the more similar the sequences of the target protein the better the homology model is likely to be. However, these models are of more value if there is experimental evidence to validate their use. When a particular target is inaccessible to structural biology, a project may rely on the use of a related protein for structure determinations. In many cases, a surrogate may simply be an orthologous protein from another species or a similar member of the same gene family. Sometimes a surrogate may exhibit similarity at the structural and functional level that does not extend to sequence, for example in the case of thermolysin and NEP.

ADMET Modeling

Potency and simple property filters such as the rule-of-five are often the main criteria in the lead discovery stage of a drug. To design a candidate medicine from the initial lead, however, one needs to consider a host of additional parameters that can affect the biopharmaceutical and safety properties of the drug such as the *in vivo* absorption, distribution, metabolism, excretion, and toxicology (ADMET). The tools of structural bioinformatics, namely sequence-structure relationships and protein homology modeling, can be employed in the field of ADMET modeling.

The most developed work in this field is in the area of cytochrome P450 modeling to predict drug metabolism. The metabolism of a drug by various cytochrome P450 enzymes is an important factor in the development of a drug. The route of metabolism can affect the drug's half-life, dose, and even safety, since P450 polymorphisms result in differential metabolism. In the absence of a structure of a human P450 enzyme, homology models can be employed to model the P450 active sites. Combining structure-activity relationships with P450 protein homology modeling enables the production of pharmacophores that are capable of predicting compound metabolism with success.

Although the structure of no human cytochrome P450 is known, the crystal structure of the rabbit CYP2C5 was recently determined. The availability of a mammalian P450 has considerably improved many of the models and sequence alignments of the human P450s and thus should enable the constructions of more predictive pharmacophores for a range of metabolizing enzymes.

Future Directions

With the advent of the genome era and the expected increase in the availability of protein structures from the introduction of structural genomics, it is likely that in the next few years most drug discovery programs against a soluble protein will begin with a structure or model of the target protein.

The early use of structural information in a drug discovery project can provide a great deal of insight for the lead discovery program. Quantitative target assessment can play a great role in the investment decision of whether or not to pursue a project. Many medicinal chemistry projects have floundered for the simple reason that the binding pocket of the target did not have the required physicochemical properties complementary to binding a potent small molecule. Thus, initial analysis of binding site can provide a significant guide to the ultimate success of the target. A robust equivalent to the rule-of-five for molecular targets will enable automatic target assessment for the large number of novel structures expected from the structural genomics initiatives.

In additional to assessing targets, comparative structural bioinformatics holds promise in identifying new drug targets from combined study of pathogen and human genomes. In comparing active sites, computational target triaging methods can simultaneously assess target druggability and species selectivity. Identifying interspecies differences in the binding sites of drug targets can also be exploited in either the choice of *in vivo* models or by specifically engineering animal receptors by site directed mutagenesis to mimic the human receptor binding pocket. Knowledge-based approaches, combined with the current explosion in sequence and structure data, may move us to a new *prospective* paradigm in which it may be possible to discover a suitable drug against a given target long before any application is known. Combined with advances in single-nucleotide polymorphism detection, this may make possible individualized medicines in which each patient gets a drug designed against his or her particular form of the target. As we move toward a situation where drug discovery projects are bathed in structural and sequence information, it is the role of the structural bioinformatician to integrate this wealth of data accelerating drug discovery.

9

MOLECULAR VISUALIZATION

A clear, concise visual representation of a macromolecular structure, be it a single image, a pair of stereo images, or a full-blown, interactive three-dimensional view, remains probably the most eloquent way to describe the very significant volume of data that is encapsulated in the atomic coordinates of a model. The goal of this chapter is to examine the different ways in which macromolecular structures may be represented, and to give brief overviews of a few of the macromolecular visualization packages that are currently available.

The programs and packages mentioned here are some of the most commonly used examples of macromolecular visualization software and they can be divided roughly into three classes: the first and possibly the least *visually* demanding visualization task surrounding macromolecular models is the construction of an atomic model. The starting point for model building may be a blank screen and a sequence, in which case the model must be built largely from scratch, or it may be an existing structure that can be modified and molded to fit the data for the target structure. Either way, the process is highly interactive, and packages that deal with this specialized area of visualization are generally large and complex but at the same time rather limited in the styles of representation that are available, the emphasis being on the data themselves rather than their representation.

Once a structure is built, refined, and available for wider use, the challenge becomes that of obtaining useful information from the structure. In many cases simply the shape and secondary structure composition of a structure can be invaluable, but an atomic resolution model contains dramatically more information than just this. Extracting detailed information from a model requires tools that allow interactive manipulation and query of atomic coordinates, from measuring distances and angles to displaying a series of overlaid multiple structures. Although there are more possibilities for using different styles of representation at this stage, the emphasis at this point is again on clarity and visual simplicity, and more importance is generally attached to tools that enable a user to interrogate a structure than to those that provide a range of visual styles.

Finally, once the structure is ready for publication, software is required to generate clear, informative, and attractive representations of atomic data, most often in the form of static images, but possibly also as two-dimensional animations or three-dimensional, interactive scenes. Displaying a macromolecular structure on a high-powered graphics workstation allows one or perhaps several users to interactively explore and investigate that structure, but unfortunately the mass of information that is easily gathered by a single user can be difficult to condense sufficiently to make it amenable to wider distribution. Presentation software must filter the large volumes of data into a form suitable for the low-detail mediums that are currently used to distribute structure information, generally still a two-dimensional figure in a standard printed journal.

Those researchers trying to determine the first molecular structures were faced with not only the theoretical challenge of calculating electron density values, but also the daunting task of somehow interpreting this electron density and obtaining the atomic coordinates that constitute a theoretical model of a macromolecular structure. The earliest molecular models, such as that of myoglobin, were built from masses of rods, wires, and spheres, so complex that the molecule itself was often lost in the web of supporting metalwork that was required to maintain its structure.

The technical challenge of actually constructing such models from electron density was eventually met by the Richards Box, a construction affectionately known as Fred's Folly. These elegant, if cumbersome, optical devices involved stacks of glass or Perspex sheets onto which electron density contours were traced out by hand.

A half-silvered mirror was used to superimpose an image of the electron density contours on an image of the physical model, so that the operator could see the model overlaid with the experimentally derived map. Peering into the dim image the crystallographer could manually build the model of the protein structure by joining together small fragments of molecule and adjusting them to fit the faint outlines of electron density by eye.

As with many other areas of science, macromolecular structure determination truly took off with the advent of electronic computing, and, as graphics technologies developed, so did the field of macromolecular visualization. Until relatively cheap and plentiful computing power became available, any calculation of the kind required for molecular structure determination was a painful manual undertaking, while visualization still involved either a hand-built physical model or a computer-generated, two-dimensional representation, formed by plotting electron density values or atomic coordinates on paper. Gleaning any useful information from the model or plot required good spatial awareness and not a little imagination.

The earliest attempts at electronic representations of molecular models used a computer-controlled oscilloscope to display a rotating image of a protein structure, with the speed and direction of rotation being controlled by the user. The system had many shortcomings (the image had to be constantly rotated to give any impression of three-dimensionality, the model was fixed and could not be altered by the user, and the hardware required to drive the system was itself specialized and experimental), but it was an essential proof of concept and undoubtedly a herald of things to come.

Once Levinthal and colleagues had demonstrated the power of electronic visualization, several groups constructed graphics systems of their own, using whatever number-crunching and display systems they had available locally or could build for themselves. These early graphics systems were milestones in molecular visualization, as crystallographers were finally able to view their models as truly three-dimensional objects, but they were mostly ad hoc constructions, difficult to build and almost as difficult to use. At about the same time, in the late 1960s, the head of the computer science department of the University of Utah and a professor from Harvard founded a company with the goal of developing and advancing the new field of computer graphics. In 1969 Evans and Sutherland (E & S) produced one of the first commercial vector graphics systems, the crude and very expensive Line Drawing System (LDS 1). Although it was far from a financial success, LDS 1 proved that computer graphics was an invaluable way to display complex data, and gave the company the impetus to develop a range of graphics systems and workstations that were the workhorses of the field of macromolecular visualization for many years.

Vector graphics systems were limited in the range of styles of representation that were available, and they gradually gave way to far more flexible systems that used raster-based displays. Raster systems were capable of richer and more detailed depictions of molecules than simple lines, and, as they developed, the software that was used to display molecular structures was also improved to take advantage of the new possibilities. One of the earliest and most widely used programs for constructing and manipulating molecular structures was FRODO, written by Alwyn Jones and colleagues. FRODO provided high-quality, interactive, color images of electron density maps and structures, and gave the user the ability to fit a model into displayed density by moving fragments of the model or even individual atoms.

The program also provided sophisticated tools such as a structure "*regularizer*" that could improve the geometry of a model by refining bond distances and angles against idealized values, helping to make the process of model building more accurate and less error prone. Originally written for the DEC PDP 11/40 with a Vector General 3404 display, FRODO was ported to the E & S Picture System 2 and other graphics platforms, including later E & S systems. The program remained in widespread use throughout the 1980s before eventually being superceded by O. O was intended to correct many of the architectural problems that had built up in FRODO, allowing more complex features to be added and giving more flexibility for the user. The program is still under development and remains one of the most popular crystallographic model-building packages.

At around the same time as the crystallographic community was making the switch from FRODO to O, at the other end of the spectrum in terms of complexity and features, lightweight structure viewers were also being developed, solely for the purpose of examining molecular structures. One such program was RasMol, developed by Roger Sayle as part of his graduate work in the early 1990s. The program started life as a test-bed for ideas about interactive rendering and gradually became what is now one of the most popular and widely used general-purpose molecular visualization programs. KINEMAGE was even more lightweight than RasMol, but it was designed for a somewhat different purpose: the goal of a KINEMAGE scene was to encapsulate a visual description of the molecule that was created by the author of the structure, and provide the user with views and descriptions of the model that were written by the person who knew the structure best. Although some features of KINEMAGE were limited, even compared to RasMol, its ability to display an authored scene, with labels, annotation, and specific view orientations, was undoubtedly ahead of its time and most of these features are yet to be reproduced by any of the mainstream visualization programs currently available.

The next major sea change in molecular visualization was once again the result of new developments in computer technology. By the start of the 1980s the computer science department of the University of Utah had a reputation for computer graphics, having already spawned Evans and Sutherland and given a number of young researchers a head start in the computer graphics industry. Among the graduates of the Utah computer science department was Jim Clark, who, in 1982, founded Silicon Graphics, with the aim of making the most powerful graphics platform in the world and making it affordable. By 1987 Silicon Graphics had achieved these goals and had produced the de facto standard in computer graphics. Large Silicon Graphics systems powered flight simulators for the aviation industry, powerful rendering clusters were used to create stunning special effects for movies, and Silicon Graphics desktop workstations were *the* platform for molecular visualization.

Until very recently Silicon Graphics had a virtual monopoly on scientific computers, making powerful workstations with high quality three-dimensional-graphics capabilities, and selling them at a price that individual academic groups could readily afford. However, in late 1990s the revolution in Personal Computer (PC) hardware brought major changes to all aspects of structural biology, as cheap and extremely powerful desktop PCs largely outpaced all but the most expensive workstations in both raw power and graphics capability. Driven principally by the electronic games industry, PC graphics technology has rapidly caught up with the more established graphics platforms from established vendors such as Silicon Graphics, and at the time of writing it is possible to buy a $1,000 PC that can match or even outperform a dedicated graphics workstation costing an order of magnitude more. It is already possible for any user to generate views of macromolecular structures that are both visually appealing and at the same time almost bewilderingly detailed, and, as graphics technologies continue to develop, the possibilities for macromolecular visualization can only increase.

Visualization Styles and Software

The styles used to represent a given structure change radically according to the various uses of the representation. At the outset, when trying to solve a structure by crystallography, the goal of the crystallographer is to turn a blank screen and a sequence into a three-dimensional atomic model of a structure, and representation styles necessarily focus on the constituent atoms of the molecule. NMR structure solution is radically different from crystallographic structure solution, and although some interactive manipulation of models can be required at certain stages of the process, the tools used for this manipulation are often the same as those used for electron density interpretation.

Although there are now refinement/model building packages, such as ARP/wARP, that can reliably construct crystallographic atomic models largely automatically, they still rely on having relatively high-resolution electron density maps to work with, and in most cases crystallographers still pore over visual representations of electron density and construct models using complex, interactive software packages.

Several visualization packages are used for crystallographic model building, and common to all of them are the ability to display electron density, and tools for manipulating atomic coordinates to fit that density. Probably the most widely used of these packages is *O*. It incorporates a wide range of model-building tools, which, like those of its predecessor, FRODO, are aimed at making model construction and manipulation simpler and more accurate. With a similar set of features, *XFit*, part of the XtalView crystallography package, is an alternative model-building program. In contrast to the command-driven interface of O, XFit is entirely mouse-driven, and all functions can be accessed from a graphical user interface, making it slightly easier to learn than O and less intimidating for beginners. *QUANTA* is a commercial package for macromolecular crystallography from MSI (now Accelrys) that provides a similar range of tools to the other model-building programs, but also integrates with other MSI products that can perform model refinement and simulations.

The representation styles used by all of these model-building packages are similar: electron density is invariably represented as the now-familiar three-dimensional "chicken-wire" contours, since this generally gives an observer the best impression of the three-dimensional shape of the node of density that is being interpreted without obscuring the atomic structure that is being manipulated. For the same reason, probably the most useful style for displaying atomic coordinates for manipulation is the simple wire-frame bonds representation, with the lines being colored according to the type of the atoms being linked (red for oxygen, blue for nitrogen, etc.). Basic lines can usually be drawn very quickly by most graphics systems, so even a large structural ensemble can generally be represented in this fashion without bringing the computer to a complete halt.

Other representation styles, although rarely used to represent an entire molecule, are often used to complement simple line drawings. In the case of ligands or heterogenous compounds that are bound to a larger structure, drawing the smaller molecule with atoms and bonds represented by solid spheres and rods (or "*sticks*") can be useful in helping the observer to pick it out from a complex scene. A common way to demonstrate the approximate van der Waal's radii of atoms is to draw them as large, solid spheres, also known as *space filling* or CPK representation, and this can also be useful for highlighting the position and interactions of smaller molecules bound to larger ones.

Many different programs can generate these relatively simple representations of molecules, but probably the most widely used is *RasMol*. It is simple to use, yet flexible, and can be used for display and interrogation of atomic models. Thanks to the clever programming behind its graphical interface, and unlike most full-blown structure manipulation packages such as O and XFit, RasMol can comfortably display a full, all atom representation of even large molecules in CPK or ball-and-stick style and still allow the user to manipulate the view and query the scene interactively. RasMol runs well on practically every platform, from PC/Mac through all flavors of Unix and even VMS, and is probably the most useful general-purpose structure viewer currently available. As well as the original version, there are other versions of the program that build on the basic RasMol to add various new features. *RasTop*, for example, is a Windows-only program that adds a more comprehensive user interface and extensions to the features of the original RasMol such as mouse-based selection of regions of a model, additional scripting commands, and improvements in the display options.

The next level of detail at which molecules are commonly represented utilizes somewhat abstract views of macromolecular (and in particular, protein) structures. The propensity of proteins to form well-defined secondary structural elements—α-helices and β-strands—is a fundamental property of protein structure. Jane Richardson pioneered a style of representing α-helices as simple cylinders or broad, spiral ribbons, and β-strands as broad, flat ribbons, and this remains one of the most enduring and appealing ways of representing protein secondary structure. By abstracting away the atomic coordinates and representing a structure according to secondary structure alone, this schematic style aptly describes both the arrangement of individual atoms in adjacent residues along the chain, as well as the interaction between more widely separated atoms through hydrogen bonding.

Practically all programs that can be used to view macromolecular structures can also generate interactive, Richardson-style, three-dimensional representations of protein structures. An important noninteractive program for generating attractive images of molecular structures is *MolScript*. MolScript scenes are described using a powerful scripting language and available representation styles for atomic coordinates include wire-frame, CPK, or ball-and-stick styles, while secondary structure elements may be drawn as solid spirals and arrows for α-helices and β-strands, respectively. Output formats include PostScript and JPEG, but probably the most useful output format is a simple three-dimensional representation that can be fed in to *render*, a ray-tracing program from the *Raster3D* package that can be used to produce high-resolution ray-traced images of the MolScript scene. Version 2 of MolScript adds a useful graphical front end, which allows scenes to be previewed in an interactive OpenGL viewer before they are written directly as an image or passed to an external rendering program.

A widely used modification of the original version of MolScript, known as *bobscript*, adds enhanced coloring capabilities and, most significantly, the ability to display electron density maps as solid surfaces or meshes, although there is no facility for previewing a scene as in MolScript v2.

Along with electron density, many visualization programs can display various other kinds of three-dimensional data, such as electrostatic charge. In some cases it may be appropriate to display the data in the form of meshes, as for electron density, but another common style of representation is a projection of the data values onto molecular surfaces. The earliest visualization programs, such as FRODO, were able to give an idea of the van der Waal's surface of a molecule, usually using arrays of dots plotted at the van der Waal' s radius for each atom, but as computing power and graphics capabilities have improved, many users now have access to graphics hardware that can readily display interactive representations of molecular surfaces as solid three-dimensional shapes. *Grasp* was one of the earliest programs to be able to display surfaces interactively and although it is no longer in development and is available for only one computing platform (SGI), it remains one of the most commonly used programs for looking at the properties of macromolecules. Electrostatic potential is a property that lends itself well to being mapped onto a three-dimensional surface. Grasp is capable of coloring a solid surface according to the density and polarity of charge surrounding surface vertices, and the resulting patches of positive and negative charge dramatically illustrate the nature of the charge on the surface of a molecule.

Several newer programs can also display various three-dimensional data as solid and even translucent surfaces. One such package is *Chimera*, which can display both electron density and various kinds of surface as meshes and opaque or semitransparent solid surfaces. The core features of Chimera include a flexible interactive viewer for displaying molecular structures in a wide variety of styles, as well as basic structure editing tools, but the program also has a modular design that allows new features to be added easily. External modules, written in python, a powerful high-level scripting language, add the ability to render arbitrary volume data, to perform semiautomated docking of small molecules to larger structures, and to share modeling sessions across the network between users in different physical locations. Although designed to be extensible, Chimera is essentially closed-source, and is distributed only in binary form, at least at present. In contrast, *PyMol*, another flexible, extensible package for molecular visualization, sports many of the same features as Chimera, but is freely available and open source. The program supports a wide range of representation styles, from simple line drawings through chicken wire or solid molecular surfaces and density maps. Control is via a python-based scripting language and scenes can be manipulated interactively using a built-in viewer, output directly as images, or ray-traced with another built-in module to produce publication quality images.

Finally, one of the most powerful packages that can be used for visualizing molecular surfaces is *AVS*, from Advanced Visual Systems. AVS is a completely general-purpose visualization tool that uses a novel graphical editor to design networks that link together many separate modules, each of which performs a single task. It is notoriously difficult to master, but, with the right modules and careful design of networks, it can generate good results and provides an extremely flexible environment for investigation of molecular properties and structure and for all types of interactive visualization.

WEB-BASED VISUALIZATION SOFTWARE

Historically, in order to run a particular program, a user has been forced to obtain, compile, and install that program locally before he or she can even begin to use it. With the advent of the World Wide Web and the explosion of interest in server-based rather than locally installed software, this scenario is no longer always the case. Technologies such as Java, Active X, JavaScript, and so forth mean that users can run increasingly complex applications without having to explicitly download them, and rely instead on having the application delivered automatically along with a Web page, all for a single click of a mouse.

Java applets are probably the most commonly used form of Web-deliverable, platform-independent software, allowing developers to write a single program that can be deployed on many different kinds of computer, from PC to Macintosh to Unix, without any changes to the code or data it uses. Although Java has been slow to catch on with certain platforms, it is now widely available and is fairly well supported by all of the major Web browsers on most platforms. For simple, low-resolution visualization tasks, Java is a very capable solution and is becoming popular with developers because of its power and ease of use, and with users because of the ease of deployment of applications.

WebMol is a lightweight Java applet that can display and query a molecular structure. Since it is focused mainly on interrogation of a scene rather than visualization, it can display a structure in only a few different styles, but it does provide some sophisticated tools for querying atom-level data from a model. Tools such as live, interactive Ramachandran plots make WebMol a useful tool for assessing the quality of a model and for extracting detailed information from that model. Distance-matrix viewers make WebMol a useful tool for assessing the quality of a model and for extracting detailed information from that model. In the same vein, another Java applet, *QuickPDB* is one of very few visualization programs, Web- based or otherwise, that provides tools for interrogating both the structure and sequence of a protein structure simultaneously. Although the tools are rudimentary, QuickPDB is, as the name suggests, quick and easy to use, even over slow Internet connections, and is accessible from practically every reasonably recent Web browser. QuickPDB is one of the visualization options from the Research Collaboratory for Structural Bioinformatics (RCSB) Protein Data Bank (PDB) site. The *Molecular Interactive Collaborative Environment* (MICE) is another Java applet for viewing molecular structures, but it provides less support for querying the structure and places more emphasis on the representation of the structure. The key feature of MICE that differentiates it from other lightweight structure viewers, and indeed, from even the more complex molecular visualization packages, is that MICE allows users to generate interactive views of any structure in the PDB and to then share that view in real time with other users across a network. Multiple users in different physical locations can collaborate over a single shared scene, passing control between users as needed, and using tools such as a shared pointer to identify regions of a scene that are under discussion. Generation of the scene, creation of a collaboration, and control of an ongoing collaboration are all possible through simple form interfaces, and no configuration on the part of the user is required to make use of any features, making MICE simple to use, as well as powerful. Like QuickPDB, MICE is also available as one of the visualization options in the RCSB PDB site.

Web browsers began life as simple tools for displaying simple pages, but as the potential of the browser was realized, it became clear that there was no way that developers of the browser itself could hope to keep up with the mass of applications, technologies, and services that would soon be available on the Web. Netscape got around this problem by introducing a "plug-in" architecture to their browser that would allow third-party software developers to create modular applications that would be installed inside Netscape and used to cope with suitably enhanced Web pages and sites. Microsoft Internet Explorer soon followed suit and now most major Web browsers have some mechanism for making use of third-party modules. *Chime* is a commercial derivative of RasMol from MDL Information Systems, packaged in the form of a browser plug-in but having all of the same capabilities and flexibility as its stand-alone cousin. Once installed in a Web browser, Chime is invoked to handle PDB format files, creating a RasMol-style view of the structure. As well as being useful in its own right as a means of viewing downloaded PDB files on the fly, Chime also forms the kernel of the *Protein Explorer*, a Web application that provides a framework for examining and manipulating structures, using animations and pregenerated descriptions of structures of interest.

Tool Kits for Visualization

At least in the field of molecular visualization, most applications are still written as individual, stand-alone programs or as a collection interconnected applications that possess a set of features and requirements that are determined entirely by the author. Although the developer may be receptive to the requests and suggestions of the users, changes and enhancements are usually still under the immediate control of the author of a given package. Many visualization programs are distributed only as precompiled binaries for a given platform, and others may be available as source code that is designed to build and run on only a small range of platforms. Users do not expect packages to be significantly configurable, nor to be readily extensible by anyone but an experienced programmer.

This model of software development is by far the most common, but it is also rather restrictive. The end users of a program must accept the shortcomings of the available software or they must generally start from scratch and write their own bespoke package to suit their needs, which inevitably leads to large numbers of small, unconnected programs being written by large numbers of developers working independently, each writing ad hoc, undocumented scripts or programs to solve precisely the same problems as their peers.

A better, more efficient solution is to provide users with high-level tool kits, from which they can quickly and relatively easily construct custom programs to fit any problem at hand. This model of development underlies several current visualization projects; the idea that users should be given high-level modules that each perform a specific task, and a framework within which to arrange them, rather than complete, monolithic programs with features and capabilities they will never use.

There are several of these tool-kit-style packages under development. One such is project is *Birdwash*, a set of modules dealing with building, refinement, and analysis of macromolecular structures, written by one of the authors of O, and building on the experiences of designing and implementing a complex visualization and structure manipulation package. At time of writing the Birdwash tool kit includes modules for visualizing atomic structures and electron density, and a means of editing structures as is required for crystallographic model building. Similar but unconnected projects are the *Molecular Modeling Toolkit* (MMTK), aimed more at molecular simulations rather than model construction, and the continuation of the MICE project, the *Molecular Biology Toolkit* (MBT).

The goal of MBT is to construct a pure Java tool kit that will provide high-level tools for displaying and editing both macromolecular structures and sequences. Probably the most advanced tool kits and modules are those from the group of Art Olson, that specialize in creating graphical tools for structure visualization and manipulation. Having previously developed modules for manipulating molecular structures under AVS, the group has considerable experience in the design of such tools, and their current tool kit includes a variety of python modules, including a structure loader, a surface generator, structure viewers, and molecule docking tools, all of which can be used to make purpose-built applications with only a minimum of high-level programming.

Any list of software will inevitably be biased by the interests and experience of the author. Given the wealth of programs that are currently available for visualizing macro- molecular structure and taking into account the ongoing development efforts in the field, it is also inevitable that such a list will be incomplete and out of date as soon as it is written. This chapter gives only the briefest of overviews of a few of the packages that are available to anyone wanting to look at the structures of macromolecules, but it is hoped that by illustrating just a few of the possibilities, the reader will be able to assess the usefulness and usability of these and other packages for themselves, and will be able to find the most appropriate one for any given task.

10

Mouse Genome Database

Two important resources for mouse biological and genomic data and analysis exist, and continue to expand at The Jackson Laboratory (U.S.A.). These resources, the *Mouse Genome Database* (MGD) and the *Gene Expression Database* (GXD), have as their long term goal the facilitation of research through access to integrated genomic structural data, gene expression information, and phenotypic descriptions. The *World Wide Web* (WWW) provides a method for easily navigating these integrated data resources to address complex questions of biological importance.

The goals for the MGD and GXD resources are (i) to provide high quality, elemental genomic data suitable for analysis and integration with other biological data, (ii) to develop software for data importation, analysis and display, and (iii) to develop easy-to-use flexible interfaces for access by the scientific community. Emphasis is on acquisition of primary data, since these are most suitable for analysis and combination with new data.

Mouse Genome Database (MGD)

Beginnings

The Mouse Genome Database developed as an outgrowth and extension of various information resources, both paper and electronic, on the laboratory mouse. The tradition of pre-publication and anecdotal information exchange in the mouse genetics community fostered the early development of data compilations and consensus map building.

Early compilations of mouse genetic data pre-date electronic databases. The first gene description catalog for the mouse was published from The Jackson Laboratory in 1941 by Dr. George Snell.

Dr. Margaret Green should be credited as the developer of the first mouse genetics database when, in the 1950s, she began an index card file system delineating published and personally communicated results of experimental crosses. These formed the basis of early versions of a composite mouse genetic linkage map. Later, she also compiled descriptions of mutant and polymorphic genes in the mouse, the centerpiece for the first edition of *Genetic Variants and Strains of the Laboratory Mouse* published in 1981. These gene description data (later known as the *Mouse Locus Catalog*, MLC) were maintained as a word-processing document.

In the 1980s, GBASE (*Genomic Database of the Mouse*), the first online resource of mouse genomic information, was developed by Drs. Roderick and Davisson. GBASE provided a single menu for user access to three independent data sets: Locusbase, an Ingres database with a character cell interface, contained summarized mapping data initially populated from Dr. Green's cards; MATRIX, with a command- line interface, contained strain-by-locus allele data; and MLC, with an IRX text searching interface, contained synoptic descriptions of genes.

In 1989, the first incarnation of the *Encyclopedia of the Mouse Genome* (*Encyclopedia*), a suite of software tools for viewing mouse genetic data, was developed through the collaborative work of Drs. J.H. Nadeau, L.E. Mobraaten, and J.T. Eppig. This software provided intuitive, graphical user interfaces for browsing genetic and cytogenetic maps, associated references, notes, and gene descriptions. Its purpose was to provide simultaneous access to information derived from different database sources and to provide means of querying those various sources using a single computer mouse 'click'. The *Encyclopedia* offered graphical browsing of the Chromosome Committee reports for the mouse and the MIT Genome Center SSLP maps under both UNIX and Macintosh operating systems.

Also during the 1980s, a number of domain specific databases were developed to fill specific research needs and produce periodic publications. Among these were a database containing primary haplotype mapping data to support linkage analysis and map drawing programs, a database of probes, clones, and molecular markers characterizing these new molecular reagents and associated RFLP data, and a database of homology relationships between the mouse and other mammalian species. A compilation of the characteristics of 728 laboratory mouse strains also was initiated during this period by synthesizing information from *Cancer Research* listings about inbred strains dating back to 1952.

In 1992, the Mouse Genome Database was initiated with NIH funding. Several significant challenges came with MGD ' s early development. Initial requirement analysis, database design, and implementation had to be accomplished in a backdrop of continued maintenance of the existing GBASE online resource. The pre-existing databases had to be assimilated and, most importantly, integrated. Although this legacy provided a foundation of data already in electronic form, the data existed on different platforms, in different data management systems, and with fundamentally different data paradigms and organizations. Further, these databases contained varying amounts of overlapping information (e.g., gene symbols), requiring extensive programmatic and manual comparisons, and data reconciliation, and data resolution.

Present

In 1994, the first public release of MGD appeared on the WWW, although full integration of all pre-existing systems was not yet complete. Since then, MGD has continued to evolve its underlying structure to maximize data integration, as well as expand its data coverage. Data volume has grown continuously; many new types of data sets have been added, and software developments have enhanced data access and provided new tools for users to explore data relationships. MGD is updated continuously with new data added daily. Software changes to the database schema, enhancements to the user interface, and new analysis and display tools are developed, tested and made available through our WWW site at 3-4 month intervals.

MGD includes structural genomic and phenotypic data about the mouse. Currently, the data sets include:

1. *Genetic marker characteristics*: gene identification and definition data; gene symbol, name, alleles, nomenclature
2. *Mapping data*: experimentally generated genotypic data from linkage crosses, recombinant inbred strain experiments, and many other genetic mapping techniques
3. *Physical mapping data*: probe vs. YAC hit/miss data; STS content data
4. *Comparative mapping data*: for mouse and >50 other mammalian species
5. *Maps*: genetic, cytogenetic, physical maps ofmouse; comparative maps between mouse and other mamals
6. *Molecular segments*: clone, probes, primers, ESTs; their characteristics and sequence links

7. *Polymorphisms*: visible, biochemical, RFLP or PCR variant polymorphisms among strains
8. *Phenotypic data*: descriptions of genes, mutations, and their function
9. *Inbred strains*: descriptions of inbred strains of mice, including their unique quantitative characteristics.
10. *Chromosome committee reports*: for viewing or downloading.

Data acquisition for MGD includes curation of the published literature, data downloads from genome centers, and individual researcher data submissions and annotations. Current data downloads are from a number of DNA mapping panel providers, including the EUCIB (European Collaborative Interspecific Backcross) panel, The Jackson Laboratory panel, and the NCI-Frederick Cancer Research and Development Center panel; the Whitehead Institute MIT (mouse physical maps and data); the I.M.A.G.E. (Integrated Molecular Analysis of Genomes and their Expression) consortium (clones and libraries); and the Washington University/Howard Hughes Medical Institute mouse EST project.

The WWW interface to MGD provides a variety of search options, with search results presented as summaries, tabular details, or graphical

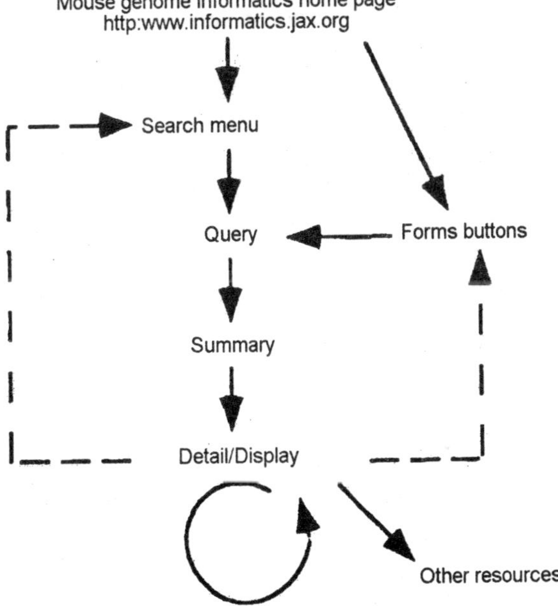

Fig. 10.1. General WWW navigation of MGD.

displays. The users' general navigation paradigm involves selecting a starting point for data query, focusing the search by optionally filling in fields in a query form, and further selecting specific data sets from those that satisfy the query. Detail data pages contain hypertext links to other data in MGD or in external data resources, where relevant, further enriching the information available to users. In addition, a number of pre-generated reports are available for frequently requested large data lists, including complete lists of genetic markers and complete tables of mouse-human and mouse-rat gene homologies.

Maps are viewed in three formats: in the browser window (Web map) with hypertext links to gene data; using the interactive *Encyclopedia of the Mouse Genome* software tool; or as a publication quality map printed from a PostScript file. Each genetic and comparative map is generated 'on-the-fly' based on user defined parameters specifying data set, marker types or classifications, region to be displayed, and whether to show homologous genes for another species. Cytogenetic and physical maps currently are only available as Web maps.

MGD strives to provide the scientific community with the most up-to-date information in an easy-to-use environment. Special requests for information not easily retrieved through the WWW interface can be accessed directly using the public SQL server. Alternatively, our User Support team will do specialized user queries on request.

Gene Expression Database (GXD)

Differential gene expression generates complex spatio-temporal networks of gene and protein interactions. The laboratory mouse, as an important animal model in the study of human disease, is being used extensively in gene expression studies. Emerging high throughput methods make it possible to analyze thousands of genes simultaneously for expression in different tissues. Such experiments will provide global expression profiles that can be used to guide focused expression studies using more conventional expression assays, such as Northern and Western blot, RNA *in situ* hybridization, and immunohistochemistry, to determine what transcripts and proteins are produced by specific genes, and where and when these products are expressed at the cellular level. The goal of the Gene Expression Database is to support the storage and analysis of all these data, with the initial concentration being mouse embryonic development.

In February 1996, the GXD Index, a view into the literature on mouse embryonic gene expression, was made available through the

WWW. This initial offering provides users with a searchable index of research reports documenting data on endogeneous gene expression during mouse development. For each scientific publication, the Index includes the genes studied, the embryonic ages analyzed, and the expression assays used. The GXD Index is integrated with MGD to foster a close link between genotype, expression, and phenotype information.

In February 1998, a 'cDNA and EST Expression' search was added. Unlike the MGD 'Molecular Probes and Segments' search, the GXD format enables better searching on source data associated with expressed sequences. Particularly, tissue, age, cell line, and library are prominent search fields.

In June 1998, the first large-scale expression data set was searchable through the 'Gene Expression Data' query form. These data, generated by T. Freeman include RT-PCR assays for 517 genes in 45 mouse tissues from 6-8 week old animals and from 15 day old embryos. Assay data include primary data on each sample prepared, expression profiles for each sample in each experimental gel, and images of the gels.

Importantly, although only limited expression data are available at this writing, GXD is poised to undergo rapid data expansion. The database is now implemented to capture and make available other datasets and several other types of expression information. Expression patterns are described by a comprehensive dictionary of anatomical terms that has been developed in collaboration with Drs. Bard and Kaufmann of the University of Edinburgh and Drs. Davidson and Baldock of the MRC, Western General Hosptial, Edinburgh. Further, editorial interfaces for capturing and updating assay data from publications and image scanning are in place.

Some expression data will be acquired from annotation of the literature by database editors. However, because only a fraction of the gene expression data that a laboratory generates actually appears in published form, it is anticipated that GXD data largely will come from electronic data submissions. The *Gene Expression Annotator* (GEA) has been developed for this purpose and is currently being tested by several laboratorics. The Annotator prototype provides important features for capturing standardized descriptions of gene expression data, validating data, and submitting data files. GEA supports a drag-and-drop facility for importation and indexing of image files, a hierarchical look-up list for embryonic anatomy, and links to resources such as MGD and GenBank for verifying nomenclature and describing probes.

With the sum of these developments, GXD will quickly expand and new data sets and features can be expected to appear regularly on the Mouse Genome Informatics WWW site.

Mouse Genome Informatics (MGI) WWW Site

MGD and GXD together provide a unique resource for analyzing how the structural genome, through developmental pathways, produces observed phenotypes. These databases are tightly integrated to enable comprehensive analysis of genotype, expression, and phenotype data. Further, as a practical consideration, because MGD already contains many data types that need to be shared with GXD (*e.g.,* data on genes, molecular probes, inbred and mutant strains), efficiency is gained by coordinating maintenance of these data.

MGD and GXD provide user access through a common interface, the Mouse Genome Informatics WWW site. This site was developed with the premise that biological scientists interested in accessing data on the genetics and biology of the laboratory mouse prefer to access data in a coordinated, integrated system. The MGI WWW site provides seamless access to MGD and GXD, with extensive hypertext linking between the data sets, such that the user can traverse all data without intentionally having to switch from a MGD to a GXD view, and vice versa. Importantly, the MGI WWW site will allow future expansion into other areas of mouse biological data, such as mouse tumor information, which is slated to be released at our site later this year.

MGD and GXD will continue to grow, expand, and evolve. Many new challenges are anticipated as both high throughput and exquisitely

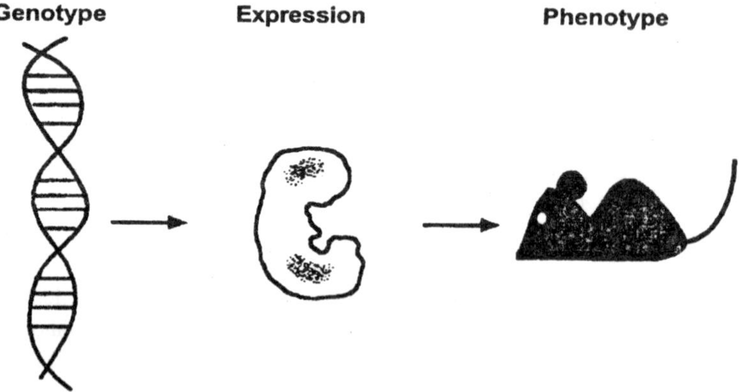

Fig. 10.2. The mouse genome database (MGD) provides genetic, genomic, and phenotypic data on the mouse.

sensitive methodologies continue to develop, allowing new and exciting biological data to be collected. As well, evolving computer technologies will give us the ability to analyze and display these data in flexible graphical ways. The mouse has become a pivotal animal model system. New emphasis is being placed on sequencing its genome and on generating new mouse mutants by homologous recombination and saturation mutagenesis. Comprehensive genetic, expression, and phenotypic analysis of these mutants will provide unprecedented insights into the mechanisms that underlie normal development and disease. MGD and GXD will continue to play an important role in the acquisition and analysis of these critical biological data for the laboratory mouse.

Basic Structure and Informatics

The rapid accumulation of information from *in situ* experiments on the developing mouse embryo, for example hybridization or immunohistochemistry, together with its spatial and temporal complexity demands a spatio-temporal database to allow collation, comparison, analysis and query. The patterns of gene-activity and consequent effects on the tissue development are alternative views of the structure of the embryo which has traditionally been described using morphologically defined structures or *anatomy*. To record this data, in order to understand the complex interactions between genes and the consequent morphogenesis and differentiation, we require a means of mapping the spatial information onto a neutral representation. The most appropriate representation for this task is the mouse embryo itself therefore we are generating 3-D image or *voxel* models of the mouse embryo, initially at each developmental stage defined by Theiler, but with the possibility of extension to finer time-steps especially at the earlier stages. The voxel models are 3-D arrays of image values corresponding to a conventional histological section as viewed under the microscope and can be digitally re-sectioned to provide new views to match any arbitrary section of an experimental embryo. The design of the *gene-expression* database, and the reconstruction methods, have been presented before and will not be repeated here. In this paper we discuss the underlying design of the spatial aspects of the atlas and database, and some of the associated bioinformatics issues.

The purpose of this database is to provide a means of comparing and analyzing gene-expression, or other spatially organized data, within the developing mouse embryo. Anatomy is our "external" description of the embryo while images of expression domains of different genes brought together into a single space-time frame provide an alternative "internal" description and a primary aim of developmental genetics is to understand the "external" description in terms of the internal gene activity. For this reason our database must include an unambiguous mapping between space and anatomy. This implies a standardized anatomical nomenclature as well as a mapping from anatomical term to spatial region or *domain* in the reconstructions. This nomenclature database has been defined and is Internet-accessible.

The combination of the 3-D voxel models, the standardised nomenclature (and associated lineage and tissue information) and the mapped anatomical domains is known as the *Edinburgh Mouse Atlas* (EMA) and is being developed as a collaboration between the Department of Anatomy, University of Edinburgh and the *MRC* Human Genetics Unit. The nomenclature is also the key element for a purely textual description of gene-expression patterns and provides the link between the spatio-temporal gene-expression database under development at the MRC, UK and the gene-expression database (GXD) at the Jackson Laboratory, USA.

Voxel Models and Database Structure

Database systems for spatially mapped information have been developed and in long use as Geographic Information Systems and a number are now emerging for biomedical research and teaching. The special feature of biomedical atlases is that there is no single correct object, i.e. the underlying structures show natural variation and the processing steps to acquire the data introduces additional systematic and random spatial variation. This raises additional problems of data mapping in order to enable spatial comparison. For gene-expression patterns the basic data arises from images of tissue sections or "whole-mounts" which provide a projection of the whole 3D pattern onto a 2D image. To map this onto a standard embryo it must be possible to find the corresponding section or projection, and then transform or warp the experimental data onto that atlas section or projection. In the future, mapping methods from 3D data, e.g. from confocal microscopy, MRI or section reconstructions, such as those developed for brain-mapping will have to be provided. The simplest reference model which can allow resectioning and mapping of spatial information is a 3D array of image values or voxels.

For the purposes of developmental genetics the voxel model should be *interpretation-free*, i.e. it must be possible to map information independently of any other interpretation or description of the embryo (e.g. anatomy). This is important because in general the visible morphological features (anatomy) at any stage will not provide adequate descriptors for the gene-activity which may be related to the development of form not yet visible. For the purposes of the database we have selected the most commonly used histological staining (H&E), and digitized images of the serial sections in order to reconstruct the 3-D image. For the EMA the spatial resolution is determined by the original microtome section thickness which is 2 microns for Theiler stages 1-12 and 6 to 8 microns for older embryos.

Experimental data is mapped onto the models as a list of voxel locations (3D binary image or *domain*). The gene-expression database will thus have a similar structure to that shown in figure 1 for the anatomy. The reference reconstructions act to define the coordinate system onto which the database entries with spatial domains can be mapped. These initially will be the anatomical components and gene-expression patterns but can be extended to any spatially organized data.

The image processing software adopted for this project has been developed by the MRC Human Genetics Unit and is known as *Woolz*. This is implemented in C and is in the public domain. The gene expression database has an object-oriented design and we have selected Objectstore (www.odi.com) as the database management system because it allows the data object to be defined in C++, which provides transparent access to the Woolz objects and procedures for the purposes of update and query.

Atlas Spatial Coordinates

In defining an atlas it is important that the coordinate system make biological sense, provide help for navigation and allow some temporal (stage to stage) comparison and measurement. The image data has its own digital coordinate system corresponding to the original section planes and discrete (voxel) coordinates which represent the basic resolution of the database. To accommodate both coordinate systems the image data is held in its original digital form and an affine transform (scaling, 3D rotation and translation) is defined which converts image coordinates to "real-space" embryo coordinates.

The real space embryo coordinates are defined by an origin and two spatial directions which define two of the axes of a right-handed Cartesian coordinate system. For the older embryos the origin is defined to be the rostral end of the notochord which also defines the direction of the z-axis. The x-axis is then aligned to point through the centre of the adjacent dorsal part of the neural tube. By this definition and at this location in the embryo, the x-axis points ventral-dorsal, the y-axis points right-left and the z-axis points caudal-rostral.

This definition will suffice for all embryos older than Theiler stage 11 when the notochord becomes visible. For earlier stages the origin and caudal direction can be defined by the primitive streak with symmetry used to define the other coordinates. For blastocysts the shape of the blastoceolomic cavity provides the required origin and directions although the rotational symmetry implies that the orientation of the x- and y-axes is arbitrary. For each reconstruction, the rule used to determine the "real" coordinate frame is accessible from the database.

ATLAS TEMPORAL COORDINATES

The time sequence of "significant" embryological events is not linear in the sense that the time resolution required to distinguish these events varies from fertilization to birth. Furthermore, it is well known that embryos of different strains develop at different rates. For this reason the use of "clock" time, for example days *post coitus* (dpc), is not always useful as well as being difficult to determine. To overcome this issue other time-scales have been defined which depend on developmental events, for example the first appearance of "fingers", or the closure of the eyes. The most widely used scheme for such *staging* is that due to Theiler. For particular periods of development other staging systems are important namely Downs and Davies (D&D) for the early embryos, and *somite count* for the period from about 8 to 10 days dpc. Experimentally determined gene-expression patterns will come from embryos that may not match the defining criteria for a particular stage because of inter-strain differences or simply natural variation and the embryos may not have been staged using the Theiler system. In this situation the data will be submitted with the authors' best estimate in whatever staging system used, and kept so that a user can be aware of how the stage was estimated. For search, that temporal information will be converted to an equivalent Theiler stage *range*, since in general there may not be sufficient precision to determine a single corresponding Theiler stage. The transformation between different staging systems is encapsulated in the table published by Bard *et al* (1998) and is reproduced on the WWW pages of the Mouse Atlas Project. To accommodate the finer grain descriptions of the development sequence provided by Downs and Davies for the early embryos, the "*integer*" Theiler stages have been extended to "*floating-point*" thereby defining sub-stages in the Theiler system.

VIEWING COORDINATES

To submit data to the database or to make queries involving spatial coordinates the user must be provided with an interface in which to define spatial position or region. The simplest interface appropriate to this task presents the user with a section view which is a cut at an arbitrary orientation through the voxel model. The *viewing coordinates* define this section, how it is displayed on the screen and the transform from screen to model coordinates for mapping data or DB query.

To define the sectioning plane two angles are required to define the viewing direction and a perpendicular distance from a fixed point will define the section. For the viewing angles we define*pitch* as the angle by which the view is tilted away from the vertical, and *yaw* as the rotation of that view around the vertical. These two angles are the usual spherical coordinates, θ and ϕ and define the normal to the viewing plane. The other parameters that define the plane are the fixed point f and distance d.

The transformation from the image coordinate $r = (x,y,z)^T$ to section coordinate $r' = (x',y',z')^T$ is given by

$$r' = \mathrm{R}(r-f), \text{ with } z' = d,$$

and the rotation matrix R is defined in terms of rotation by the three Eulerian angles ξ, η, ζ (xsi, eta, zeta) by

$$\mathrm{R} = \mathrm{R}_\xi \mathrm{R}_\eta \mathrm{R}_\zeta \text{ where}$$

$$R_\xi = \begin{pmatrix} \cos(\xi) & \sin(\xi) & 0 \\ -\sin(\xi) & \cos(\xi) & 0 \\ 0 & 0 & 1 \end{pmatrix},$$

$$R_\eta = \begin{pmatrix} \cos(\eta) & 0 & -\sin(\eta) \\ 0 & 1 & 0 \\ \sin(\eta) & 0 & \cos(\eta) \end{pmatrix} \text{ and}$$

$$R_\zeta = \begin{pmatrix} \cos(\zeta) & \sin(\zeta) & 0 \\ -\sin(\zeta) & \cos(\zeta) & 0 \\ 0 & 0 & 1 \end{pmatrix}.$$

The two Eulerian angles ξ and η are equal to pitch and yaw respectively and with the fixed point and distance determine the section plane.

To determine the transform from screen to voxel coordinates the third Euler angle must be fixed. The choice of this angle depends on how the user wishes to view the data and we define two viewing schemes termed the "up-is-up" and "statue" modes. For "up-is-up" the user defines an up-vector within the 3D reconstruction so that the projection of that vector onto the selected plane will be parallel to the displayed y-axis. By default this vector is defined to point up through the head so that a coronal or sagittal view will be displayed with the head neural tissue at the top of the image. For this mode the third Eulerian angle zeta is calculated by finding the component of the up vector u that is perpendicular to the viewing direction v, i.e.

$$w = u - \alpha v, \text{ where } v = R\begin{pmatrix} 0 \\ 0 \\ 1 \end{pmatrix}$$

and R is the rotation matrix for $(\xi, \eta, \zeta) = (\theta, \phi, 0)$. This vector is in the original coordinate frame and the required angle is determined by transforming w to the viewing coordinates and calculating the angle with respect to the y' axis:

$$\zeta = -\tan^{-1}\left(\frac{w'_x}{w'_y}\right) \text{ where } w' = Rw.$$

The "statue" mode can be understood by considering the voxel image to be a statue with the user walking around on a horizontal plane i.e. the x-y plane of the image. The section seen by the user is rotated about its axis of intersection with the horizontal for display on the screen. For the user the effect is that rotating a section view around a vertical axis will result in a gradual rotation of the displayed image. The Euler angles for this viewing mode are $(\xi, \eta, \zeta) = (\theta, \phi, -\theta)$.

NAVIGATION

Locating a specific section within a 3D volume can be quite difficult so, in addition to the option of panning though to a recognizable position which can then be defined as the fixed point, a number other navigational aids are provided. The first is the option of interactively defining a second fixed point which then reduces the number of degrees of freedom to one, namely rotation around the line joining the two points. This is termed the ***torsion***, denoted τ, and uniquely defines the viewing direction

$$n = \cos(\tau)n_2 + \sin(\tau)n_3,$$

where the orthonormal vectors n_1 and n_2 are defined by rotating the coordinate frame so that the z-axis is parallel to the vector joining the two fixed points, namely the vector

$$n_1 = \frac{f_2 - f_1}{|f_2 - f_1|}.$$

This defines a viewing direction and (θ_0, ϕ_0) corresponding rotation matrix R_0 by $\cos(\theta_0) = n_{1z}$ and $\tan(\phi_0) = n_{1y}/n_{1x}$ and hence

$$n_2 = R_0 \begin{pmatrix} 1 \\ 0 \\ 0 \end{pmatrix} \text{and } n_3 = R_0 \begin{pmatrix} 0 \\ 1 \\ 0 \end{pmatrix}$$

The new viewing angles (θ, ϕ) are then given by $\cos(\theta) = nz$ and $\tan(\phi) = n_y/n_x$, and are parametrised by τ, which ranges from 0 to 2π.

The second means of navigation is to use *fiducial* points,i.e.points that have been previously defined and exist in the Atlas database for the purposes of navigation. Selecting a single point defines a rotation center, two points will define a line as above and a third will fully define a plane.

Other navigational support is provided by a 3D feedback window in which the section plane is displayed relative to the image bounding box and selected anatomical structures.

Data Submission

The Edinburgh Mouse Atlas has been developed primarily to support a gene- expression database by providing a spatial framework for the mapping expression and anatomical information. The methods for mapping spatially organized data onto the reconstructions can be categorized under the headings: text, painting or warping.

Text Submission

The spatial domain is described by listing anatomical components or existing database entries. Typically the user will define a domain by selecting anatomical components with the assumption the required domain is the union of the component domains (which by definition do not intersect). If in addition the user wishes to use predefined gene-expression domains these can be used to refine the new domain by using the set operations *union*, *intersection* and *difference* on the corresponding sets of voxel locations.

Painting

An alternative mean of defining the domain is for the user to delineate or paint the new domain directly onto the reconstruction using the Edinburgh Mouse Atlas paint program or equivalent software. This allows the user to select an arbitrary section and delineate regions directly using a variety of painting "tools". This program also allows the editing of domains defined by any other means therefore can be used in conjunction with a text definition of the domain to allow additional refinement. This is likely to be more efficient than defining the whole 3D domain *de novo*.

Warping

Submission using text and/or painting can be time consuming and most graphical data will be entered by using image processing both to extract the required pattern and to transform or *warp* the data onto the atlas. Before generating the submission it is assumed that the user has digitized the required experimental data, then the simplest means is to locate the matching section from the atlas and for the user to define "tie-points" that can be used to produce a warp transformation, e.g. as a thin-plate spline or polynomial, to map the experimental data onto the atlas image. If the resultant transformation is not correct then the user can add additional tie-points or directly edit the resulting domain using painting. There are many possibilities for determining the warp transform including a fully automatic matching. In practice the mechanism must be fast enough to allow interactive adjustment and editing.

So far we have considered section data which assumes that the experimental section corresponds to a planar section through the atlas embryo. If the experimental data is in the form of a 3D image then full 3D warping will be necessary e.g.

Data Extensions

The initial implementation of the image-mapped database will store the domains (binary image) of gene-expression only. This means that expression strength information can only be encoded as separate domains, e.g. weak, medium and strong, or by storing a dithered version of the grey-level image. The Woolz image format includes a simple extension to grey-level types within the image domain and therefore we plan to extend the schema to allow full grey-level images of gene expression data to be submitted. This means that the query language can be extended to use grey-level operations such as gradient and intensity.

11

AGRICULTURAL GENOME

The *Agricultural Genome Information System* (AGIS) provides Internet access to genome information from agriculturally important organisms. The server delivers information from thirty-six databases, encompassing mostly crop and livestock animal species, including the databases for all of the major food crop genome projects. Also included are a number of databases which have related information, such as databases for several model organism genome projects (including ACeDB for the nematode, *Caenorhabditis* elegans, and DictyDB for soil amoebae, *Dictyostelium discoidium*), reference databases such as Mendel (plant gene nomenclature), PhytochemDB (plant phytochemicals), EthnobotDB (plant uses), Fire Ant, and OMIA (Online Mendelian Inheritance in Animals), as well as links to other important resources like AGRICOLA.

Table 11.1. AGIS databases

Plant Genome

- AAtDB—*Arabidopsis*
- Alfagenes—alfalfa (*Medicago sativa*)
- BeanGenes—*Phaseolus* and *Vigna*
- Cabbagepatch— *Brassica*
- Chlamy DB--*Chlamydomonas reinhardtii*
- CoolGenes—cool season food legumes
- CottonDB—*Gossypium hirsutum*
- GrainGenes—wheat, barley, rye and relatives
- MaizeDB—maize
- MilletGenes—pearl millet
- RiceGenes—rice

RoseDB—Rosaceae
SolGenes—Solanaceae
SorghumDB—Sorghum bicolor
Soy Base—soybeans
TreeGene—foresttrees
Mendel—plant-wide gene names

Livestock Animal Genome
BovGB ASE—Bovine
ChickGBASE—poultry
PiGB ASE—swine
SheepBASE—sheep
OMIA—Online Mendelian Inheritance in Animals

Other Organisms Genome
ACeDB—C. elegans
DictyDB—The soil amoebae *Dictyostelium discoideum*
MycDB—Mycobacteria
PathoGenes—fungal pathogens of small-grain cereals
RiceBlastDB—the rice blast fungus *Magnaporthe grisea*

Plant Reference
AGRICOLA—plant genetics subset
CIMMYT—Wheat International Nursery Data
Ecosys—plant ecological ranges
EthnobotDB—worldwide plant uses
FoodplantDB—Native American food plants
MPNADB—medicinal plants of Native America
PhytochemDB—plant chemicals
PVP—Plant Variety Protection
PVPSoy—Soybean Plant Variety Protection Data

Insect Reference
Fire Ant —*Solenopsis*
Face Fly—*Musca autumnalis*
Horn *Fly—Haematobia*
Screwworm—Cochliomyia hominivorax
Stable Fly —*Stomoxys calcitrans*

AGIS uses the World-Wide Web (WWW) technology to distribute information. Any person with a forms-capable browser can access the databases. The forms interface was kept as simple as possible, using only standard HTML commands, so as to maintain compatibility with as many WWW browsers as possible. All current versions of Netscape browsers work, as does Internet Explorer and even old versions of Mosaic.

Implementation

ACEDB, the genome database developed for use with the *C. elegans* genome project, is used as the back-end. While other commercial database systems were initially considered, ACEDB is the *defacto* standard for genome projects. Thus, by using ACEDB, data compatibility with the AGIS collaborators was maintained.

Genome data delivery from AGIS has evolved, keeping pace with new computer technologies. Originally, information was distributed on the Internet by a Gopher server, and by CD-ROM for researchers who had no Internet access. Both methods were discontinued as AGIS was migrated to the World-Wide Web by the use of a highly modified ACEDB program, which generated static HTML pages for text displays and GIF images for graphical maps. The current, more interactive WWW interface is provided by a package called webace, which utilizes forms-based HTML pages to provide more access to ACEDB functions, such as the Query by Example and Table-maker facilities, as well as interactive GIF images.

The databases are incorporated into the aceserver layer, which runs as an inetd daemon using RPC calls. The aceserver is a modified form of ACEDB, and includes the giface program, which turns ACEDB graphics into GIF images suitable for use by webace.

The webace program is a PERL script that runs as a CGI program, utilizing the CGI.pm PERL module. Webace translates user queries into the ACEDB query language, and then converts ACEDB objects into HTML documents for display. HTML links are created by a set of simple markup rules, and can insert URL anchors which point to other ACEDB objects, external database queries, or even external analysis programs like the NCSA Biology Workbench.

Guided Tour of the AGIS Databases

From the main AGIS, menu, the user can access the list of available databases by following the "Databases" link. Databases are grouped according to type (plant, livestock, model organism, and reference). Following each of the database listing is a set of links (browse, query, about), which allow the user to jump to the particular HTML forms set to access data from that particular database.

There are two basic access methods for the plant genome data in AGIS: (1) Browsing interface; (2) and Query interface. Each method has advantages and disadvantages, and the interface of choice will be strongly dependent upon both the data desired and the familiarity of the user with ACEDB. Database access is restricted to single databases at present — no concurrent or cross-database querying is supported. The following sections present the three interfaces arranged by complexity and power. Finally, a section on the hypertext ACEDB objects is presented.

Browse Mode

For novice users, the browse mode is certainly the simplest interface. Browse is a point and click interface which allows the user to wander through the data in a completely hyper-linked mode. A problem with this approach is the complexity and amount of the data can be overwhelming. Still, using the browse mode allows the user to avoid the complexities of the ACEDB query language.

Selecting the browse mode presents the user with a form listing the available classes in the database. Selection of a class (in this case Locus) presents the user with a list of all objects available for that class. When there are too many objects to be comfortably listed, the list is collapsed into a set of sublists. Selecting a sublist brings up a shortened list of all objects in that particular range. The user always has the option to import the entire list.

Query Mode

Selecting query mode brings the user to a form allowing the databases to be searched/queried by six different methods, including fuzzy search, WAIS, Query by Example, Query Builder, Table-maker, and ACEDB Query Language in which an ACEDB query can be input directly.

ACEDB query

This mode is extremely fast and efficient for data retrieval, but it does require the user to be familiar with the ACEDB query language as well as the data structure of the particular database. While these two topics are outside the scope of the current paper, the reader is referred to several excellent treatises on the ACEDB query language, available at the AGIS site.

Fuzzy (AGREP) and WAIS query

The fuzzy and WAIS modes present the simplest query interface. One or more words can be entered as the search string, and wildcards are accepted. A search returns a hypertext list of database objects.

Query by example

This mode allows the user to query one class of a database by typing search strings into one or more field categories on a form. The query brings back a list of matching objects. Figure 6 shows the Query by Example interface to the GeneFamily class of the Mendel database.

Query builder

The Query Builder interface allows much more complex queries than Query by Example, permitting the user to string together any number of "and", "or", and "xor" queries together for a particular class from a database.

Table-maker mode

The most powerful method of accessing data from ACEDB data is the Table-maker program. Table-maker allows the user to create a relational database style table of ACEDB objects. Like the query interfaces, Table-maker requires an understanding of the structure of the database. However, the challenge of acquiring this knowledge is more than made up for in improved information retrieval.

Table-maker works by making an initial query, then applying modifiers and performing "follow" operations, which can be thought of as automated linking operations between objects.

Hypertext ACEDB Objects

Regardless of the mechanism used to initially query the database, an ACEDB object is the ultimate goal of interacting with the databases. The object looks very similar to a native ACEDB object, with links to other objects represented as hyperlinks. Following such a link will replace the current object by following the link.

One salient difference between native ACEDB objects and those presented by the AGIS system is that while ACEDB objects launch multiple viewing types (e.g., maps, text, images), the default type of any object returned by the AGIS system is text. At the top of each object is the "View as Graphic" link, which invokes the GIFace server and returns a clickable ISMAP graphic. While mildly annoying to those looking specifically for graphical views, the text orientation of the AGIS server was a conscious decision designed to conserve bandwidth and avoid burdening users who have slower network connections with unwanted graphics.

New technologies bring new opportunities to improve service, and the providers of AGIS have always striven to make use of as much state-of-the-art technology as possible. One such new technology is the

Java programming language which allows interactive viewers and programs to be downloaded across a WWW link. JAVA holds promise to improve the AGIS interface, allowing it to be much more interactive and responsive than permitted by standard HTML forms.

Jade is the Java version of ACEDB. A new experimental AGIS server has been created, with small modifications to the basic Jade structure to give it a familiar look and feel. While the basic interactions are similar to the WWW version, the Java map viewers are a great improvement. Additionally, the Java version for the first time holds forth the promise of concurrent database queries and cross-species comparisons performed directly from the AGIS servers.

Maize Genome Database

In 1923, R. A. Emerson addressed a detailed letter to "Students of Corn Genetics', soliciting community solutions to issues of gene nomenclature. Building on this stimulus, Emerson hosted a 'cornfab' in his hotel room, during the December 1928 Genetics meetings in NY. April 1929, Emerson and colleagues disseminated the first Maize Newsletter (MNL): a mimeographed summary of the 'cornfab', along with a list of 20 available stocks, a list of curators for individual linkage groups, a summary of available linkage information, and 78 references where linkage data were to be found.

The first published linkage map compilation in 1935 showed 62 loci. In 1991, USDA-ARS initiated a plant genome database project, and tasked the editor of the MNL, Ed Coe, to develop a maize genome database. The 1991 MNL, now volume 65, included 1439 Stocks available from a USDA-ARS funded Stock Center, 840 entries on the Gene List with 423 key references, an additional 776 references from the annual literature and some 950 colleague addresses. Data from this issue were transferred into the Fall 1991 prototype MaizeDB.

The 1991 prototype was based on insight gained by the Coli Genetics Stock Center, New Haven, CT with an industry standard software, Sybase, for the database management system, and the development of Genera software to create forms for query and data entry and define entities without writing code. This choice of software has permitted concentrating resources on data curation by a team of 'biologists', with support of a systems analyst.

The Sybase software had been employed by various other species genome databases, including the human genome. A sample of the Genera form specification for the short locus query form is appended to this chapter, along with the form specified, and a sample query returned. Indexed flat-files which provide full-text searching have been a standard feature of the Genera software. In 1994, MaizeDB, employed Genera upgraded for WWW form and full-text query access; the upgrade also permitted facile linking with external databases. Data relevant to the maize genome exists in major sequence, reference and germplasm databases, as well as other species-specific genome databases, including *E. coli*, yeast and other plant genomes. Reciprocal, record-to-record links with SwissProt were established June 1994, and followed soon thereafter with reciprocal links to GRIN, the Germplasm Resource Information Network for the US.

Data Content

MaizeDB maintains the current genetic maps for the community, together with supporting documentation, and information about gene function and expression. Supporting documentation may includes (1) data analyses summaries, such as QTL experiments; (2) raw data, map scores and recombination data; (3) references, with address information for many of the authors; (4) access to research tools, including genetic stocks, DNA clones and PCR primer sequences. Gene function and expression are provided in comments on the locus pages, in the gene product, mutant phenotype, trait information about mapped and unmapped. Gene function and expression are provided in comments on the locus records, by links to gene product, mutant phenotype, and agronomic traits.

Over 110,000 records are currently maintained by MaizeDB; major classes of information are summarized in the below table. There may be upwards of some 40,000 genes in maize. Of the 15,764 maize loci in the database, there are 6,028 genes, where 1,188 have been mapped to chromosome arm or better by mutant phenotype or single copy probe. The MNL Gene List, now includes 1,512 unique genetic factors of which 827 have some map position. Loci have expanded from the 62 phenotypically identified genes on the 1935 map compilation to include pseudogenes, probed sites (PCR or RFLP); restriction fragments (mitochondrial maps), chromosomal segments, points and QTL (*quantitative trait loci*).

The bins map strategy has been employed by the MNL editor to unify the extensive maps in the community, and without misrepresenting known order. Order of loci within a bin may be ascertained by examining the empirically determined maps, which are provided within the database.

The MaizeDB server also hosts the electronic MNL, abstracts for the annual meetings, and provides a public clearing house for gene nomenclature and registry of new names.

Database Design and Features

The database contains 27 entities, 340 tables, 53 views, 50 triggers and 221 stored procedures. Stored procedures written at Missouri generate an ACeDB product that was first released in June 1993. Major enhancements to the 1991 prototype were implement March 1993; QTL experiment representations were added 1994. Minor design upgrades, or example merging two entities complete with data, creating a Trait entity, and writing triggers have been accomplished with the core MaizeDB staff at Missouri. Schema may be found at our public ftp server; both computer and biological documentation about fields, tables and entities may be accessed by WWW form query.

The database supports controlled vocabularies, with a thesaurus. All major entities have synonym tables. The most carefully curated terms are the entity "Type" terms, those defining relationships between loci, and the keywords used for reference annotation. The data entry forms permit selecting from the controlled vocabulary, and also entry of new terms if appropriate. In contrast to many of the controlled vocabularies, for example body parts, terms for entity types are kept to a minimum, and the use of broader and narrower terms avoided. Instead, there are tables for entity properties, where a locus of type Gene or Cytological Structure may share a common property, for example MNL Gene List.

External Database Links

Over 13,500 records have over 62,000 links to some 30 external databases. When more than one record links to the same external database record, each link is counted in the summation. Links to loci are most often indirect, and linked directly to a gene product, probe, variation or reference. While only 2% of the loci may be linked to an external database, 10% of the locus variations have links, as do 27% of Probes, 32 % of Gene Products, 30% of the Stocks and 18% of the References, excluding the MNL and maize meeting abstract links.

Links to external databases are provided by two modes: user request for data from the other database, where typically a particular database may be selected for retrieval of a sequence, or by MaizeDB pre-determined choice or 'jump'. In both cases, external databases are treated as Persons. The code for the 'jump' is embedded in Genera and utilized in MaizeDB both for Person and for Loci. An example of user-choice for external database:

DB Key	*DB*	*Variation*
M61191	DDBJ	wxl-B6
M61191	EMBL	wxl-B6
M61191	Entrez	wxl-B6
M61191	GenoBase	wxl-B6
M61191	GSDB	wxl-B6

An example of a jump is provided by the maize locus wxl, which lists several related loci as below:

Relation	*Locus*
Left marker	W11 white 11
Right marker	d3 dwarf plant3
Contained in	Dp9 duplication 9
orthologous, putative	xWx Triticum aestivum
orthologous, putative	WX Oryza sativa

Clicking on one of the orthologous rice or wheat loci retrieves the Plant Genome database for that record.

Data Entry

Form entry is provided by the Genera software utilized for non-WWW access to the database. A small group of both on-site and off-site curators have access to the central Sybase tables using this software. Other data is entered using customized scripts, developed by curators for large electronic notebooks in the community, or standard script, such as that developed in 1993 for journal article import from standard reference manager format. It has been upgraded by MaizeDB staff to enter books and book chapters, and also MNL articles. The reference loading software matches to previously entered references, fills in missing information and reports actions taken and ambiguities.

Dissemination

WWW form permit queries based on multiple attributes, and return lists of objects that match the selected constraints. A sample form for Locus is appended. Full text query offers an opportunity to the novice user to obtain a sense for how the data are represented. Flat-files for the full-text searching are computed monthly, but retrieve real-time data.

Special Data Formats

In addition to form queries, based on selection and/or 'fill-in-the-blanks' query constraints, MaizeDB supplies several browser oriented formats on the WWW. These formats may be on-the-fly, with some or no user constraint options or they may be periodically extracted, with or without curator intervention. Hyper-links on lists will retrieve the current record from the database, based on the accession ID# in MaizeDB for the entity. Some of the lists are computed in real-time, others extracted periodically, either automated, the Core Marker list, or with curator intervention (the MNL Genelist).

Computed in real-time — no user constraints

Clicking at the Illionois Sotck Center page will initiate a real-teim query for the current listing of Stocks, either total or the 1998 additions. The catalog returned is extracted from the Stock and the Stock#Description tables and has hypertext links to the Stock entity. It provides the Stock Center accession, and a descriptive name of the Stock, as computed daily from a list of the variations associated with the Stock.

ACC	*Descriptive name*
301A	bif3-N2354
302AA	d1-N446
302AB	d1-N339
306F	ref1-MS1185
307A	Sdw2-N1991
309A	al-m3::Ds Sh2
309B	al-ml-5718::dSpm
309C	al-ml-5719A1::dSpm
309D	al-ml-5719A1::dSpm; Mod Prl

Real-time — with user constraints

Example 1. Map Scores by the BIN: user selects (1) the bin range, (2) the Mapping Panel of Stocks, and (3) format of product returned (hypertext, or comma-delimited). Data for this table are read from the Locus#Coordinates, and MapScores tables. Map Scores for loci between bins 1.01 and 1.12 using Mapping Panel 57244

Example 2. Formatted Person address. Constraint options are the city and or the person's last name. Data may be returned in a choice of 3 formats: (a) address only, (b) address and phone, as required for FedEX, or with (c) address, Email, phone and fax. Data for the below example are read from the Person, Person#PhoneNos and Persone# EMailAdresses tables.

Computed periodically, no curator intervention

Table of Core Marker information is automatically computed each week. The table content was defined by the UMC RFLP laboratory. Core markers are loci that define the edges of the bins on the consensus map; this map is curated by Ed Coe, at Missouri, with MaizeDB staff assistance.

Data are read into the output table from the Locus#Coordinates, Probe and Probe#Comments tables, as restricted by the a property 'Core Marker', stored in the Locus#Properties and Probe#Properties tables as two independent controlled vocabularies.

Periodically computed, — coordinated with curator action

The printed 1993 MNL Gene List was the first Gene List extracted from the database. The Gene List is a complex table, with the symbol, the bin location, the full name, a brief comment, putative or confirmed gene products and key references for selected loci. Hypertext links are provided to the MaizeDB Locus, Gene Product and Reference entities. Curatorial review: (1) checks that all appropriate loci are included, by examining the list of excluded maize loci of type="Gene" where there is no '*' associated with the name; (2) ascertains that the full name, the brief comment, any gene product(s) and appropriate references are updated; (3) monitors comments entered into the database for errors, completeness and currency. Creating the Gene List requires reading data into 3 de novo tables (genes, geneP, MNLGeneRefs) from several MaizeDB tables (Locus, Locus#Coordinates, Locus#Comments, Locus#GeneProducts, Anything#References, References). The only constraints are the Locus Property, MNL GeneList and the reference Annotations, First Report or Gene List. Two sample rows.:

bt1, bin(s) 5.04, *brittle endosperm1,* mature kernel collapsed, angular, often translucent and brittle (alleles sh3, sh5), may encode amyloplast adenylate translocator.

pl1, bin(s) 6.04, *purple plantl,* Pl1 plant tissues have light-independent pigment, pl1 blue light-dependent; Pl1-Bh1, colored patches in c1 aleurone and in plant; transcriptional activator for flavonoid genes; SSR phi03 1, nc009, nc010.

Dissemination to External Databases

Linking information is provided to SwissProt, to GRIN, and to EMBL, and to the Entrez-Genome division associated with GenBank.

Mapping coordinates are placed into our public ftp file, and the Entrez-Genome division is notified when there is an update. The AceDB format for the database is submitted to the central server for Plant Genome Databases at the National Agricultural library.

The Plant Genome Database suite maintained by the USDA-ARS, and others both in the US and international locations are looking towards combining sequence computation with graphical displays of data return that can represent both intra- and inter-specific data. Inter-specific genome and germplasm queries will be greatly enhanced by access to common controlled vocabularies, and metabolic databases, both inter-plant and inter-all-genomes. At MaizeDB, we anticipate enhancing user- access to the data using menu-driven, user-defined table constructions. Instead of retrieving a list of loci with tassel phenotypes, one might specify a list of the loci, any PCR primers for nearby sites, gene products, GenBank accessions, the map bins. In the near future, we anticipate an onslaught of highly structured data from enhanced funding for crop plant genomes, both US and international.

12

OPTIMIZATION

Optimization involves computing parameter values or making decisions such that certain performance or quality indices are maximized, or, conversely, some penalty or loss functions are minimized. Optimization is an area of scientific research, and also has extensive applications in technology, engineering design, etc. Optimization problems involve a large variety of situations, such as optimizing the parameters of engineering structures, developing optimal policies in decision-making problems, and optimizing controls in dynamical systems. The application of optimization techniques can also involve fitting models to data by optimizing the model parameters in such a way that output of the developed mathematical model is as close as possible to the measurements.

Optimization methods are traditionally divided into static and dynamic methods. Static optimization is the computing of extremal points of functions of one or several variables. This involves formulating appropriate optimality conditions and then proposing algorithms, often iterative, for satisfying these conditions. Dynamic optimization is the computing of optimal control functions or making optimal decisions in a sequential order, following the dynamics of some system. Dynamic optimization can involve continuous-time or discrete-time dynamics, and there are again two main approaches. One approach is to use variational type conditions leading to two-point boundary value problems. The other uses Bellman's optimality principle, which in discrete cases leads to a recursion for the cumulative score index, and its continuous-time limit leads to partial differential equation formulation called the Bellman–Jacobi equation. Here we focus on discrete dynamic optimization and dynamic programming. There are some very well known examples of application of dynamic programming in bioinformatics, namely the NeedlemanWunsch and SmithWaterman algorithms for DNA alignment.

We also include a section on combinatorial optimization in this chapter. By combinatorial optimization we mean problems which involve optimizing over paths in the plane or space, topologies of trees, graphs etc. Combinatorial optimization is related to theory of algorithms and their complexity, computer science and operation research. Combinatorial optimization is of great importance in bioinformatics because it involves a lot of non-numerical data and the need to perform various operations on these data.

STATIC OPTIMIZATION

The simplest example of static optimization is maximizing a function of one variable. The function $f(x)$ has its maximal value f_{max} at $x = x_{max}$. A characteristic of extremal points is that, provided the function is smooth, the derivative becomes zero, i.e.,

$$\frac{df(x)}{dx} = 0, \qquad ...(1)$$

which can be used for computing x_{max} by solving the algebraic equation (eq.1). At points where the derivative is negative the function decreases, and at points where the derivative is positive the function increases. This can be used for constructing recursive estimators for extremal

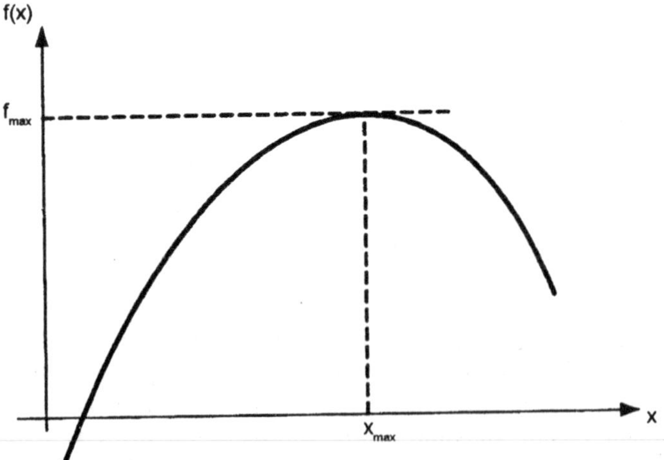

Fig. 12.1. A function of one variable $f(x)$ has a maximum f_{max} at $x=x_{max}$.

points. The condition (eq.1) applies for both maximal and minimal points. Resolving the distinction between a maximum and a minimum can be done by use of the second derivative (if the function is twice differentiable); namely if (5.1) holds at $x = x^*$ and

$$\frac{d^2 f(x)}{dx^2} > 0, \qquad ...(2)$$

then x^* corresponds to a local minimum, and if the inequality sign in (eq.2) is opposite then x^* corresponds to a local maximum.

The elementary conditions discussed above can be generalized to multidimensional functions $f: R^n \to R$. In the two-dimensional case, where the function is $f(x_1, x_2)$, the condition for an extremum analogous to (eq.1), is

$$\frac{\partial f(x_1,x_2)}{\partial x_1} = 0, \frac{\partial f(x_1,x_2)}{\partial x_2} = 0. \qquad ...(3)$$

This condition means that the plane tangent to the graph of $f(x_1, x_2)$ at an extremal point is horizontal. The vector composed of partial derivatives in (eq.3) is called gradient of f and denoted ∇f:

$$\nabla f(x_1,x_2) = \begin{bmatrix} \dfrac{\partial f(x_1,x_2)}{\partial x_1} \\ \dfrac{\partial f(x_1,x_2)}{\partial x_2} \end{bmatrix}. \qquad ...(4)$$

The notation $\nabla f(x) = \partial f/\partial x$ is often also used, for the gradient vector, where x is a vector argument.

$$x = \begin{bmatrix} x_1 \\ x_2 \end{bmatrix}. \qquad ...(5)$$

This notation is particularly useful when we need to form vectors of partial derivatives of functions with respect to subsets of their arguments. Such situations will be encountered below. Using (eq.4) and the vector notation for x_1, x_2, we can write the condition for an extremum (eq.3) as

$$\nabla f(x) = \frac{\partial f(x)}{\partial x} = 0, \qquad ...(6)$$

where there is a two-dimensional zero vector on the right hand side.

Resolving the distinction between a maximum and a minimum can be done by use of the Hessian matrix

$$Hf(x) = Hf(x_1, x_2) = \begin{bmatrix} \dfrac{\partial^2 f(x_1,(x_2)}{\partial x_1^2} & \dfrac{\partial^2 f(x_1,x_2)}{\partial x_1 \partial x_2} \\ \dfrac{\partial^2 f(x_1,x_2)}{\partial x_1 \partial x_2} & \dfrac{\partial^2 f(x_1,x_2)}{\partial x_2^2} \end{bmatrix}. \qquad \ldots(7)$$

If (eq.6) holds at $x = x^*$ and $Hf(x)$ given by (eq.7) is positive definite then the function $f(x)$ has its local minimum at $x = x^*$, and if Hessian matrix $Hf(x)$ given by (eq.7) is negative definite at $x = x^*$ then x^* corresponds to a local maximum.

Level sets (curves for two dimensions, surfaces for three dimensions, etc.) are sets for which the function $f(x_1,x_2)$ has a constant value, i.e., $\{x_1,x_2 : f(x_1,x_2) = C = const\}$. Every level set can be associated with a the constant value C of the function. When constructing algorithms for recursive maximization we aim to design a sequence of points, which "*climbs uphill*"; each point in the sequence belongs to a level line with a higher C than the previous one. If a point x belongs to a level line (or surface or hypersurface in more than two dimensions) $f(x) = C$, then which direction should we head in to increase C? This can be resolved by using the property that the gradient vector $\nabla f(x)$ is perpendicular to the level set of the function $f(x)$ and points in the direction of increase of C. This property is

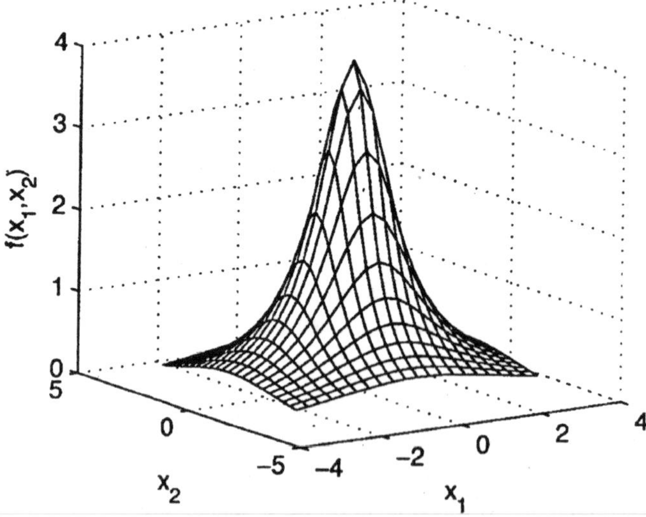

Fig. 12.2. 3D plot of an example function.

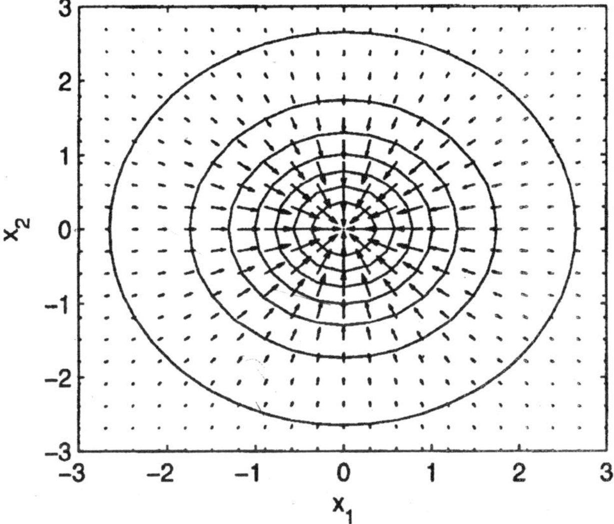

Fig. 12.3. Level curves and gradient vectors.

illustrated in Fig. 12.2, where a 3D plot of an exemplary function is drawn in the upper part and corresponding level sets and gradient vectors are depicted in the lower part. So, in many optimization procedures, the direction for the update of recursions is parallel to the gradient vector, possibly with some scaling factor, for function maximization, and the direction is antiparallel to the gradient vector for function minimization.

In (eq.3)–(eq.7), we assumed a two-dimensional space of vectors x. However, all of the above can be extended in an obvious way to vectors of higher dimensionality. Namely, for a function $f: R^n \to R$ of $x - [x_1, x_2, \ldots, x_n]^T$ the gradient vector ∇f is

$$\nabla f(x) = \begin{bmatrix} \frac{\partial f(x)}{\partial x_1} \\ \frac{\partial f(x)}{\partial x_2} \\ \vdots \\ \frac{\partial f(x)}{\partial x_2} \end{bmatrix},$$

and its Hessian matrix is given by

$$Hf(x) = \begin{bmatrix} \frac{\partial^2 f(x)}{\partial x_1^2} & \frac{\partial^2 f(x)}{\partial x_1 \partial x_2} & \cdots & \frac{\partial^2 f(x)}{\partial x_1 \partial xn} \\ \frac{\partial^2 f(x)}{\partial x_2 \partial x_1} & \frac{\partial^2 f(x)}{\partial x_2^2} & \cdots & \frac{\partial^2 f(x)}{\partial x_2 \partial xn} \\ \vdots & \vdots & \ddots & \vdots \\ \frac{\partial^2 f(x)}{\partial x_n \partial x_1} & \frac{\partial^2 f(x)}{\partial x_n \partial x_2} & \cdots & \frac{\partial^2 f(x)}{\partial x_n^2} \end{bmatrix}. \quad \ldots(8)$$

Convexity and Concavity

Convexity and concavity are notions playing an important role in many fields of mathematical modeling and among other things, in both static and dynamic optimization. By checking the convexity or concavity of a function one can distinguish between minima and maxima, as mentioned above, and turn necessary conditions for optimality into sufficient conditions.

A set $X \subset R^n$ is convex if the fact that two points x_A and x_B, satisfy, $x_A \in X$ and $x_B \in X$ implies that the whole segment with ends x_A and x_B belongs to X, which can be expressed as

$$\forall_{p \in [0,1]} px_A + (1-p)x_B \in X. \quad \ldots(9)$$

A function f defined over a convex set, $f: X \to R$, is convex if

$$\forall_{p \in [0,1]} f[px_A + (1-p)x_B] \leq pf(x_A) + (1-p)f(x_B). \quad \ldots(10)$$

A function g defined over a convex set, $g: X \to R$, is concave if it satisfies the inequality converse to (eq.10),

$$\forall_{p \in [0,1]} g[px_A + (1-p)x_B] \leq pg(x_A) + (1-p)f(x_B). \quad \ldots(11)$$

where again x_A and x_B belong to X. If $f(x)$ is convex then $-f(x)$ is concave and vice versa.

If X and Y are two convex sets then their intersection $X \cap Y$ is a convex set. If $f(x)$ is a convex function $f: X \to R$, then the set bounded by the level hypersurface $Z_C = \{x : f(x) \leq C\}$, for every value of the constant C, is convex. If $g(x)$ is a concave function, $g: X \to R$, then the set bounded by the level hypersurface $Z_C = \{x : g(x) \geq C\}$ is again convex. Owing to the property of convexity of intersections of convex sets, the sets defined by

$$\begin{aligned} X = \{x : & f_1(x) \leq C_1, \\ & f_2(x) \leq C_2, \\ & \vdots \\ & f_k(x) \leq C_k\} \end{aligned}$$

and

$$X = \{x : g_1(x) \geq C_1,$$
$$g_2(x) \geq C_2,$$
$$\vdots$$
$$g_l(x) \geq C_l\}$$

where $f_1,\ldots, f_k$ are all convex functions and $g_1,\ldots, g_l$ are all concave, are convex.

For smooth functions, convexity and concavity can be verified by use of the second derivatives. Namely, if $f(x)$ has continuous second partial derivatives in some convex set X, then $f(x)$ is convex in X if and only if its Hessian matrix (eq.8) is positive semi-definite. If $g(x)$ has continuous second partial derivatives in some convex set X then $f(x)$ is concave if and only if its Hessian matrix (eq.8) is negative semi-definite.

The linear function

$$f(x) = a^T x + c,$$

where a is a parameter vector and c is a constant, is both convex and concave. If a function $g(x)$ is convex (or concave) then $g(x) + a^T x + c$ is also convex (or concave, respectively).

Constrained Optimization with Equality Constraints

Very often one needs to optimize a function $f(x)$ with an additional condition that x belongs to some set, for example a the set of points in the plane satisfying an equation $g(x) = 0$ for some $R^2 \to R$ function $g(.)$. This is called a constrained optimization task, and can be stated more formally as

$$\min f(x) \qquad \ldots(12)$$

with the constraint

$$g(x) = 0, \qquad \ldots(13)$$

for example. It will help us to illustrate the constrained optimization problem (eq.12)–(eq.13) graphically, as we have done for a two-dimensional $x = [x_1\ x_2]^T$. In this figure, the family of level lines of the function $f(x)$ is a family of ovals centered around the same point, and the constraint curve $g(x) = 0$ is depicted in bold. When we try to minimize the function $f(x)$ we aim at moving downhill—as close to the center of the family of ovals as possible. This is limited by the requirement that $g(x) = 0$ and we see that, at the optimal point $x^* = [x_1^*\ x_2^*]^T$ the curves $f(x) = C$ and $g(x) = 0$ are tangential. The

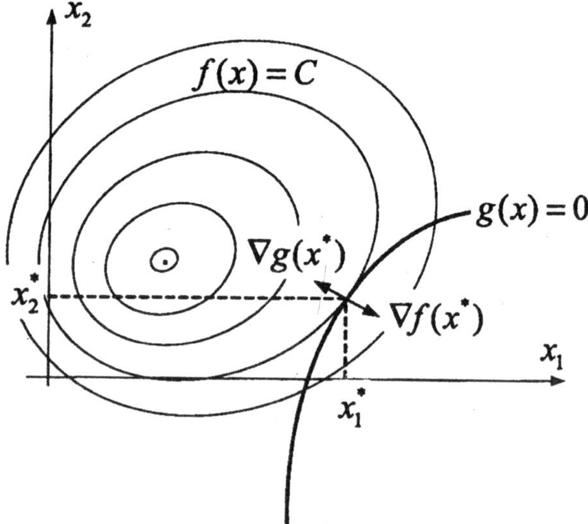

Fig. 12.4. Graphical illustration of the solution to the constrained optimization problem min f(x) with the constraint g(x) = 0, $x = [x_1, x_2]^T$.

tangentiality condition can also be understood meaning that gradient vectors $\nabla f(x^*)$ and $\nabla g(x^*)$ are parallel to each other, and so there must be a scalar number $-\lambda$ such that

$$\nabla f(x^*) = -\lambda \nabla g(x^*). \qquad \text{...(14)}$$

We have used a minus sign here to fit to with the usual notation. The number λ is called a Lagrange multiplier, and by associating with the constrained problem (eq.12)–(eq.13) the Lagrange function

$$L(x, \lambda) = f(x) + \lambda g(x), \qquad \text{...(15)}$$

we can express (14) at $x = x^*$ as

$$\frac{\partial}{\partial x} L(x, \lambda) = 0. \qquad \text{...(16)}$$

We note that computing the coordinates of the minimum x^* involves solving for three variables, namely the two coordinates of x and the scalar λ. The vector condition (eq.16) gives two algebraic equations, and the third equation is the constraint (eq.13). Equation (eq.13) can be written equivalently as a condition on the gradient of the Langrange function (eq.15) with respect to λ,

$$\frac{\partial}{\partial \lambda} L(x, \lambda) = 0. \qquad \text{...(17)}$$

One can also derive the optimality conditions (eq.16) and (eq.17) by algebraic manipulations without invoking a geometric interpretation.

The situation depicted in Fig. 12.4 involving a two-dimensional argument space can be generalized to spaces of higher dimensionality and to a number of constraint equations of more than one. This leads to the following vector formulation analogous to (eq.12)(eq.17).

Lagrange multiplier theorem

Consider the constrained problem

$$\min f(x), \qquad \ldots(18)$$

with the constraint

$$g(x) = 0, \qquad \ldots(19)$$

where f and g are functions that assign to a vector $x \in R^n$, respectively a scalar and an m-dimensional vector, respectively, i.e., $f: R^n \to R$ and $g: R^n \to R^k$. The necessary conditions for optimality in (18)(19) are

$$\frac{\partial}{\partial x} L(x, \lambda) = 0. \qquad \ldots(20)$$

and

$$\frac{\partial}{\partial \lambda} L(x, \lambda) = 0. \qquad \ldots(21)$$

where $L(x, \lambda)$ is the Lagrange function associated with the problem (eq.18), (eq.19), given by

$$L(x, \lambda) = f(x) + \lambda^T g(x) \qquad \ldots(22)$$

$$= f(x) + \sum_{i=1}^{n} \lambda_i g_i(x) \qquad \ldots(23)$$

with a vector of Lagrange multipliers $\lambda \in R^m$. In (eq.18) and (eq.19), vector notation has been used i.e.

$$x = \begin{bmatrix} x_1 \\ x_2 \\ \vdots \\ x_n \end{bmatrix}, g(x) = \begin{bmatrix} g_1(x) \\ g_2(x) \\ \vdots \\ g_m(x) \end{bmatrix}, \lambda = \begin{bmatrix} \lambda_1 \\ \lambda_2 \\ \vdots \\ \lambda_m \end{bmatrix}; \qquad \ldots(24)$$

λ^T stands for the row vector resulting from transposition of λ in (eq. 24), and $\lambda^T g(x)$ is a scalar product given by

$$\lambda^T g(x) = \sum_{i=1}^{m} \lambda_i g_i(x).$$

Constrained Optimization with Inequality Constraints

Consider the optimization problem

$$\max f(x), \qquad \ldots(25)$$

where $x \in R^n$ and $f: R^n \to R$ and with constraints of the form

$$g(x) \geq 0, \quad \text{...(26)}$$

where $g: R^n \to R^k$. The inequality in (eq.26) is componentwise; (eq.26) is a shortened vector notation for

$$\begin{aligned} g_1(x) &\geq 0, \\ g_2(x) &\geq 0, \\ &\vdots \\ g_m(x) &\geq 0. \end{aligned} \quad \text{...(27)}$$

If $x \in R^n$ fulfills (eq.27), then each of the component inequalities in (eq.27) can be either active or inactive. Inequality number i is active if $g_i(x) = 0$, and inactive if $g_i(x) > 0$. In a some sense, the problem (eq.25), (eq.26) is more general than (eq.18), (eq. 19) since each equality constraint $g_i(x) = 0$ can be represented as two opposite-sign inequalities, $g_i(x) \geq 0$, $-g_i(x) \geq 0$. The optimality problem (eq.25), (eq.26) can be resolved by use of the Kuhn–Tucker theorem.

Kuhn–Tucker theorem

We define the Lagrange function corresponding to the constrained optimization problem (eq.25), (eq.26), by

$$L(x,\lambda) = f(x) + \lambda^T g(x) \quad \text{...(28)}$$

$$= f(x) + \sum_{i=1}^{n} \lambda_i g_i(x) \quad \text{...(29)}$$

where λ is a vector of Lagrange multipliers $\lambda \in R^m$. The notation in the above is the same as in (eq.22). The Kuhn–Tucker theorem formulates the following necessary conditions for optimality in (eq.25), (eq.26):

$$\frac{\partial}{\partial x} L(x,\lambda) = 0 \quad \text{...(30)}$$

$$\frac{\partial}{\partial \lambda} L(x,\lambda) \geq 0, \quad \text{...(31)}$$

$$\lambda \geq 0, \quad \text{...(32)}$$

and

$$\lambda^T g(x) = 0. \quad \text{...(33)}$$

We shall briefly discuss the idea behind their construction. As previously, let us denote the optimal point by x^*. The condition (eq.31) is a repetition of (eq.26). The equality (eq.33) is called *complementarity* condition. The components of the vector of Lagrange multipliers which correspond

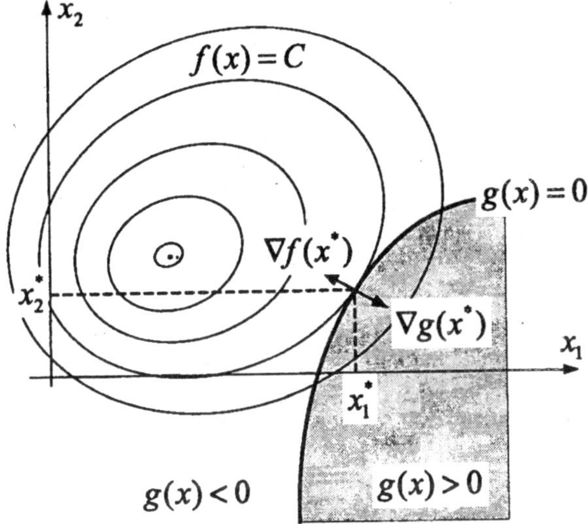

Fig. 12.5. Graphical illustration of the Kuhn-Tucker necessary optimality conditions for max f(x) subject to the constraint $g(x) \geq 0$.

to inequalities that are inactive at x^* are set to zero, i.e., $\lambda_i = 0$ and components corresponding to active inequalities are made strictly greater than zero, i.e., $\lambda_k > 0$. One can see that with this (and only with this) choice, the complementarity condition is satisfied. If one knew which of the constraints (eq.27) were active and which were inactive, at the optimal point x^*, then (eq.30)-(eq.33) would reduce to solving a system of equalities and to checking (eq.32) at found solutions. The need to add the condition (eq.32) to (eq.30) can be explained geometrically. The gradient vector $\nabla f(x^*)$ must point in the opposite direction to $\nabla g(x^*)$. Otherwise it would be possible to increase $f(x)$ and $g(x)$ simultaneously, which would contradict the optimality of the point x^*. More generally, if we denote by $i1^a > 0$, $i2^a > 0 \ldots i^a_K$ the indices of active constraints Lagrange multipliers corresponding to active constraints, then

$$\lambda_{i_1^a} \nabla g_{i_1^a}(x^*) + \lambda_{i_2^a} \nabla g_{i_2^a}(x^*) + \ldots + \lambda_{i_K^a} \nabla g_{i_K^a}(x^*) \qquad \ldots(5.34)$$

is a cone of feasible directions, i.e., the directions along which we can move the argument x without violating the constraints. If moving x along any of the directions in (eq.34) were to increase the value of the function $f(x)$, then this would contradict the optimality of x^*.

As stated above, the Kuhn–Tucker conditions are only necessary; they can be satisfied both at local and at global extrema of functions.

Situations where these necessary conditions become sufficient can be identified by employing the concepts of convexity and concavity, as presented in the next subsection.

Sufficiency of Optimality Conditions for Constrained Problems

If a convex function attains a local minimum over convex sets then, this is also the global minimum. Analogously, if a concave function attains a local maximum over convex sets, then this is also the global maximum. These properties are used to formulate the sufficiency of optimality conditions.

If, in optimization problem (eq.25), (eq.26) the scoring function $f(x)$ is strictly concave and all constraint functions $g_1(x), \ldots, g_m(x)$ are concave then the KuhnTucker conditions (eq.30)(eq.33) become both necessary and sufficient.

Consider a problem analogous to (eq.25)–(eq.26) with maximization replaced by minimization,

$$\min f(x) \qquad \ldots(35)$$

subject to constraints

$$g(x) \leq 0. \qquad \ldots(36)$$

Kuhn Tucker necessary optimality conditions stated with the use of the Lagrange function (eq.28),

$$\frac{\partial}{\partial x} L(x, \lambda) = 0, \qquad \ldots(37)$$

$$\frac{\partial}{\partial \lambda} L(x, \lambda) \leq 0, \qquad \ldots(38)$$

$$\lambda \leq 0 \qquad \ldots(39)$$

and

$$\lambda^T g(x) = 0, \qquad \ldots(40)$$

become both necessary and sufficient if the function $f(x)$ is convex and all component functions $g_1(x), \ldots, g_m(x)$ in $g : R^n \rightarrow R^k$ are convex.

Computing Solutions to Optimization Problems

For many cases of optimization problems analytical solutions leading to very useful results can be obtained by using optimality conditions discussed above. In some situations, however, no analytical expressions for optimal points are available, but proofs can be carried out of the existence and/or uniqueness of optimal points and of recursive methods which converge to optimal points. Some of these will be discussed in the following.

Some functions often appearing in optimization problems are linear and quadratic forms. A function $f: R^n \to R$

$$f(x) = a^T x, \qquad ...(41)$$

where a is a constant n-dimensional vector is called a linear form. Its gradient is $\nabla f(x) = a$ and its Hessian matrix is the zero matrix $Hf(x) = 0$.

A function $f: R^n \to R$,

$$f(x) = x^T Q x + a^T x, \qquad ...(42)$$

where Q is a symmetric $n \times n$ matrix is called a quadratic form. The common symmetry assumption of $Q = Q^T$ is due to the decomposition of a matrix into symmetric and antisymmetric components $A = (1/2)(A+A^T) + (1/2)(A-A^T)$. Only the symmetric component $(1/2)(A + A^T)$ will contribute to the value of the quadratic form $x^T A x$. The gradient vector of the quadratic form (eq.42) is

$$\nabla f(x) = 2Qx + a, \qquad ...(43)$$

and the Hessian matrix is given by $Hf(x) = 2Q$.

By comparing the gradient (eq.43) with zero we obtain the following (given that the matrix Q is invertible):

$$x^* = -\frac{1}{2} Q^{-1} a, \qquad ...(44)$$

which is a unique maximal point, provided that Q is negative definite, and a unique minimal point provided that Q is positive definite.

Simple linear regression by least squares

As an example of the application of (eq.44) let us consider the problem of fitting the parameters a and b of a straight line $y = ax + b$ to some measured data points (x_i, y_i), $i = 1, \ldots, n$. With the notation

$$y = \begin{bmatrix} y_1 \\ y_2 \\ \vdots \\ y_n \end{bmatrix}, Y = \begin{bmatrix} x_1 & 1 \\ x_2 & 2 \\ \vdots & \vdots \\ x_n & 1 \end{bmatrix}, p = \begin{bmatrix} a \\ b \end{bmatrix}$$

we can represent the sum of the squared errors of the model versus the data, as follows:

$$\sum_{i=1}^{n} (y_i - ax_i - b)^2 = (y - Yp)^T (y - Yp)$$

$$= y^T y - 2y^T Yp + p^T Y^T Yp.$$

The last expression is a quadratic form $f(p)$ given in (eq.42), where $a = Y^T y$, and $Q = Y^T Y$, and so from (eq.44), the optimal parameter fit is

$$p^* = \begin{bmatrix} a^* \\ b^* \end{bmatrix} = (Y^T Y)^{-1} Y^T y.$$

Constrained optimization problems

Here we analyze some examples, involving linear and quadratic forms, illustrating the Lagrange multiplier and the KuhnTucker constraint optimality conditions. As the first example consider minimization in $x \in R^n$

$$\min x^T Qx \qquad \text{...(45)}$$

subject to the linear constraint

$$a^T x = c. \qquad \text{...(46)}$$

Notation for a, c and Q is the same as above in this section. Lagrangian function for (eq.45)-(eq.46) is

$$L(x, \lambda) = x^T Qx + \lambda(a^T x - c),$$

where λ is a scalar Lagrange multiplier and using (eq.20)-(eq.21) we compute

$$x^* = \frac{c}{a^T Q^{-1} a} Q^{-1} a. \qquad \text{...(47)}$$

Knowing that (eq.20) and (eq.21) are only necessary conditions we recall the remarks from previous section to find whether there is maximum or minimum at x^*. If Q is a positive definite matrix then x^* is indeed a minimum of (eq.45) subject to (eq.46). If Q is negative definite, then a minimum does not exist, function in (eq.45) can approach $-\infty$ for some sequences of x, all satisfying (eq.46). For negative definite Q, a unique solution to the problem (eq.45), (eq.46) would exist if minimization were replaced by maximization.

As the second example, let us consider the following:

$$\max x_1^2 + x_2^2 \qquad \text{...(48)}$$

subject to the constraints

$$x_1 \leq 1 \qquad \text{...(49)}$$

and

$$x_2 \leq 2. \qquad \text{...(50)}$$

Here the level curves of the function (eq.48) are circles centered at $(x_1, x_2) = (0,0)$, the sets defined by constraints (eq.49) and (eq.50) are half planes, and the global maximum does not exist, in the sense that (eq.48) can be increased to arbitrarily large values without violating (eq.49) and (eq.50). Yet the Kuhn–Tucker conditions are satisfied at three points, $(x_1^*, x_2^*) = (1,0)$, $(x_1^*, x_2^*) = (0,2)$, and $(x_1^*, x_2^*) = (1,2)$. The sufficiency conditions are not satisfied, since we are maximizing a convex, not a concave function.

Finally, if we replace maximization in the above by minimization and we change directions of inequalities i.e., if we consider

$$\min x_1^2 + x_2^2 \qquad ...(51)$$

subject to the constraints

$$x_1 \geq 1 \qquad ...(52)$$

and

$$x_2 \geq 2, \qquad ...(53)$$

then the unique solution to the Kuhn–Tucker conditions is $(x_1^*, x_2^*) = (1,2)$. Here the sufficiency conditions are satisfied.

Both of the above problems (eq.48)(eq.53) are easily interpreted by using plots of functions and constraint sets in the plane (x_1, x_2).

Linear Programming

Linear programming is a special optimization problem, of finding the extremal value of a linear form over a set defined by linear inequalities. It can be formulated as follows:

$$\min a^T x \qquad ...(54)$$

subject to

$$Bx \leq b. \qquad ...(55)$$

In the above expressions (eq.54) and (eq.55), $x \in R^n$, a is an n-dimensional vector of the parameters of the linear form $a^T x$, B is an $m \times n$-dimensional matrix, b is an m-dimensional vector. The inequalities in (eq.55) are understood componentwise. The set in R^n defined by the system of inequalities (eq.55) is a (possibly unbounded or degenerate) convex hyperpolyhedron, and the problem defined by (eq.54) and (eq.55) can be understood as looking for the vertex of the polyhedron located farthest away along the direction defined by the vector a.

The formulation (eq.54) and (eq.55) is the most general in the sense that any linear programming problem can be transformed to it by introducing suitable definitions. Minimization can be changed to maximization by taking $a = -a_1$. Equality constraints can be represented by pairs of inequality constraints. There are algorithms and computer software that allow one to solve linear programming problems with very large sizes of the vector x.

Quadratic Programming

The quadratic programming problem is defined as follows:

$$\min(x^T Qx + a^T x) \qquad \ldots(56)$$

subject to

$$Bx \leq b. \qquad \ldots(57)$$

In (eq.56) Q is a symmetric, positive definite (or positive semidefinite) $n \times n$ matrix. Other parameters, a, B and b have the same meanings as those in (eq.54) and (eq.55). If the matrix Q is positive definite the problem (eq.56)-(eq.57) has a unique solution.

The optimization problem (eq.56)-(eq.57) can be efficiently solved for vectors x of large size by use of appropriate algorithms and related computer software.

Recursive Optimization Algorithms

In general, optimality conditions can be difficult to find solutions for. A solution may not exist owing either to contradictory constraints or the possibility of the value of $f(x)$ diverging to infinity. There may exist multiple solutions, even infinitely many or uncountable sets of solutions. The Kuhn–Tucker conditions (eq.30)–(eq.33), which involve both equalities and inequalities, are more difficult to find solutions for than systems of algebraic equations occuring in the Lagrange multiplier theorem (eq.20)–(eq.21); the difficult problem may be identifying the active and inactive constraints.

Even for unconstrained problems, computing optimal points is very often not possible analytically. Therefore numerical algorithms, where the value of the function to be optimized is improved step by step, are very useful and are very often applied. Below, we briefly describe some commonly applied versions of iterative unconstrained optimization algorithms.

Search for extremum of a function without derivatives

The information about the direction of increase or decrease of a function is contained in its gradient vector. However, sometimes computing the gradient vector of a function is time-expensive or cumbersome. It is therefore worth mentioning algorithms which seek an extremum recursively solely on the basis of values of the function $f(x)$, without derivatives. In the one-dimensional case, searching for an extremum of a function in an interval (x_{min}, x_{max}) can be accomplished on the basis of its successive division into smaller parts can be applied, such as bisection, or golden-section based on Fibonacci proportions. One algorithm without derivatives designed for multidimensional cases is named NelderMead, moving-simplex or moving-amoeba method, Owing to the idea behind its construction, where the aim is to localize the extremal argument of a function inside a simplex and then, recursively, shrink the diameter of the simplex to zero. Despite slow or even problematic convergence, especially in higher numbers of dimensions, it can be very useful and it is included in most software packages for optimization.

Let us assume that our aim is to minimize a function $f(x)$ in n dimensions. The algorithm needs the following parameters to be specified: ρ (reflection), χ (expansion), γ (contraction) and σ (shrinkage). The choice of values typically applied is $\rho = 1$, $\chi = 2$, $\gamma = 0.5$, and $\sigma = 0.5$. We shall describe one iteration of the algorithm, which starts from a simplex in R^n with $n+1$ vertices $x_1, x_2, \ldots, x_{n+1}$. Assume that the values of the function are ordered such that $f(x_1) < f(x_2), \ldots, < f(x_{n+1})$. Call vertices $x_1, x_2, \ldots, x_n$ the base of the simplex and the vertex x_{n+1} the peak of the simplex. First, we reflect the peak with respect to the center of the base $\bar{x}$, where

$$\bar{x} = \frac{1}{n}\sum_{i=1}^{n} x_i,$$

using the assumed value of the reflection parameter ρ. The result of this operation is denoted x_R and is given by

$$x_R = \bar{x} + \rho(\bar{x} - x_{n+1}).$$

Now, depending on the relations between $f(x_R)$ and the values of f at the vertices of the simplex, we perform different operations.

1. If $f(x_1) < f(x_R) < f(x_n)$, replace x_{n+1} by x_R and terminate the iteration.
2. If $f(x_R) < f(x_1)$, calculate the expansion point x_E, given by
$$x_E = \bar{x} + \chi(x_R - \bar{x}),$$
evaluate $f(x_E)$ and replace x_{n+1} by x_E if $f(x_E) < f(x_R)$ or by x_R if $f(x_E) > f(x_R)$. Terminate the iteration.
3. If $f(x_n) < f(x_R) < f(x_{n+1})$ compute the outside contraction point x_C, where
$$x_C = \bar{x} + \gamma(x_R - \bar{x}),$$
and evaluate $f(x_C)$. If $f(x_C) < f(x_R)$ replace x_{n+1} by x_C and terminate the iteration. If $f(x_C) > f(x_R)$, perform a shrink operation and terminate the iteration.
4. If $f(x_R) > f(x_{n+1})$, compute the inside contraction point x_{CC}, where
$$x_{CC} = \bar{x} + \gamma(\bar{x} - x_{n+1}),$$

and evaluate $f(x_{CC})$. If $f(x_{CC}) < f(x_{n+1})$, replace x_{n+1} by x_{CC} and terminate the iteration. If $f(x_{CC}) > f(x_{n+1})$, perform a shrink operation and terminate the iteration.

The shrink operation is defined as follows. Replace the vertices $x_1, x_2, \ldots, x_{n+1}$ of the simplex by new vertices $x_1, x'_2, \ldots, x_{n+1}$, where

$$x' = x_1 + \sigma(x_i - x_1),\ i = 2,3,\ldots,n+1.$$

Counterexamples can be constructed such that the above algorithm will not reach a minimal point despite the convexity of the function $f(x)$. Nevertheless, it is successful in many typical examples and, as stated, can be very useful. At least, it can be tried as a first choice.

Gradient algorithms

If we want to minimize a function $f(x)$ in n dimensions, then at a given point x_k, the direction of its fastest decrease (steepest descent) is opposite to the gradient vector $\nabla f(x)$. So we can plan a step of a minimization algorithm as follows:

$$x_{k+1} = x_k - \gamma \nabla f(x_k), \qquad \ldots(58)$$

where γ is a suitably defined parameter. The main problem is tuning the step size, adjusted by the parameter γ. If it is too small, convergence to minimum is very slow. Values that are too large will typically cause instabilities in the algorithm. Therefore there are many more or less heuristic modifications of the recursion in (eq.58), aimed at producing algorithms that both have a high speed of convergence and are robust to instabilities.

Algorithms using second derivatives

Let us start with the one-dimensional case, where a function $f(x)$ is minimized over a scalar x. We assume that the kth iteration of the minimization procedure hits a point x_k in the close vicinity of the optimal argument x^*, and that the function $f(x)$ is twice continuously differentiable, so the following approximate equation holds, since $f'(x^*) = 0$:

$$f(x_k) = f(x^*) + \frac{1}{2} f''(x^*)(x_k - x^*)^2. \qquad \ldots(59)$$

Now we think of $f(x^*)$ and $f''(x^*)$ as constant parameters, which allows us to compute the following expression for $f'(x_k)$,

$$f'(x_k) = f''(x^*)(x_k - x^*), \qquad (5.60)$$

and for $f'(x_k)$,

$$f''(x_k) = f''(x^*). \qquad \ldots(61)$$

Using (eq.60) and (eq.61), we can compute x^* from

$$x^* = x_k - \frac{f'(x_k)}{f''(x_k)}. \quad \text{...(62)}$$

Since (eq.59) is only an approximation, we take (eq.62) as the new value in the recursion rather than as a final solution, i.e.,

$$x_{k+1} = x_k - \frac{f'(x_k)}{f''(x_k)}. \quad \text{...(63)}$$

The above is called the Newton–Raphson or Gauss–Newton iterative minimum search for the minimum. An analogous derivation applies in the multidimensional case, where $x \in R^n$. If $\nabla f(x_k)$ is the gradient vector of the function $f(x)$ taken at the point of kth iteration, and $Hf(x_k)$ is its Hessian matrix, then the next iteration of minimum search algorithm will is

$$x_{k+1} = x_k - [Hf(x_k)]^{-1} \nabla f(x_k). \quad \text{...(64)}$$

In the close vicinity of the minimum, the convergence of (eq.63) or (eq.64) is very fast, but if the scheme is started from a random point it can easily end up in instability. Therefore, again scaling is commonly applied to the steps of the algorithm:

$$x_{k+1} = x_k - \gamma[Hf(x_k)]^{-1} \nabla f(x_k), \quad \text{...(65)}$$

where γ is a suitable parameter.

DYNAMIC PROGRAMMING

Dynamic programming is solving optimization problems by organizing the optimizing decisions in the sequential order. The method of dynamic programming has been applied efficiently to large variety of problems. Sometimes formulating a dynamic programming solution to an optimization problems can be tricky, and may need research. We list some properties of discrete dynamic optimization problems. Knowing them should help one to develop a suitable formulation of dynamic programming algorithm. (1) We should be able to decompose the optimization problem into separate decisions and organize the decision-making process into stages. (2) At each stage of decision-making process, we should be able to define a state of the system (a problem) which summarizes the influence of the decisions already made. (3) The scoring index should be expressed such that it can be computed iteratively, stage by stage, and provided optimal value of score for (i + 1)th stage is known we can find a recursion for ith stage. Developing the recursion in (3) is a basic technique for constructing algorithms for solving discrete dynamic optimization problems. The principle behind the recursive update of the scoring index is the following Bellman's optimality principle, *An optimal policy has the property that whatever the initial state and initial decision are, the remaining decisions must constitute an optimal policy with regard to the state resulting from the first decision.*

The recursion following from the above principle is called Bellman's equation. Below, we illustrate the principle by going through some examples, both general and more specific.

Dynamic Programming Algorithm for a Discrete-Time System

Consider a discrete-time dynamical system written in the fairly general form

$$x_{k+1} = f_k(x_k, u_k), \qquad \text{...(66)}$$

where $k = 0, 1, \ldots, K$ are discrete time instants that index stages of the optimization process, u_k stands for the decision variables, f_k is a function which gives a model for the discrete-time evolution of the system and x_k is the state of the process. Knowing x_k and u_k, u_{k+1}, ... allows us to compute future states x_{k+1}, x_{k+2},... no matter what previous decisions u_{k+1}, u_{k+1}, ... were. We assume that at each time instant k possible decisions are constrained by the requirement that u_k belongs to a set

$$u_k \in U_k(x_k), \qquad \text{...(67)}$$

depending both on the state x_k and on the time instant k. The aim is to find a sequence of decisions u_k, $k = 0, 1, \ldots, K$ such that, given initial state x_0, the scoring index

$$I(x_0) = \sum_{k=1}^{K} s_k(x_k, u_k) \qquad \text{...(68)}$$

is minimized. The above formulation is fairly general, in the sense that we allow the functions f_k, feasibility sets U_k, the components of the scoring function s_k, and the numbers of components of states x_k and decisions u_k, to change in each stage of the decision-making process.

Solution

Define the optimal partial cumulative score as

$$I_k^{opt}(x_k) = \min_{u_k, u_{k+1}, \ldots, u_K} \sum_{i=k}^{K} s_i(x_i, u_i), \qquad \text{...(69)}$$

where we have listed the optimization arguments, but have skipped the constraints (eq.67) for brevity. The solution to our dynamic optimization problem is $I_0^{opt}(x_0)$; however, we cannot compute it by direct minimization, as in (eq.69), owing to the large number of optimization variables. Knowing the optimality principle, we want to organize the minimization in (eq.69) in a recursive style. We start the recursion from the last stage of the decision-making process $k = K$, which leads to

$$I_K^{opt}(x_K) = \min_{u_K \in U_K(x_K)} s_K(x_K, u_K) \qquad ...(70)$$

and

$$u_K^{opt}(x_K) = \arg \min_{u_K \in U_K(x_K)} s_K(x_K, u_K) \qquad ...(71)$$

We assume that the above minimization, involving only the last decision u_K, can be performed efficiently. Now, the optimality principle gives a recursion between the optimal partial cumulative scores $I_k^{opt}(x_k)$ and $I_{k+1}^{opt}(x_{k+1})$, i.e., Bellman's equation, in the form

$$\begin{aligned} I_k^{opt}(x_k) &= \min_{u_k \in U_k(x_k)} s_k(x_k, u_k) + I_{k+1}^{opt}(x_{k+1}) \\ &= \min_{u_k \in U_k(x_k)} s_k(x_k, u_k) + I_{k+1}^{opt}[f_k(x_k, u_k)], \qquad ...(72) \end{aligned}$$

all allows us to compute the optimal decision

$$u_k^{opt}(x_k) = \arg \min_{u_k \in U_k(x_k)} s_k(x_k, u_k) + I_{k+1}^{opt}[f_k(x_k, u_k)]. \qquad ...(73)$$

The optimization in (eq.72) again involves only one decision and is assumed to be tractable. Starting from (eq.70) and (eq.71) and repeating (eq.72) and (eq.73) recursively, we finally obtain the solution to the whole problem, $I_0^{opt}(x_0)$.

Let us acknowledge one difficulty in the above procedure. In (eq.70), we are solving not one optimization problem, but rather a whole family of problems, parametrized by the values of x_K. Similarly, in (eq.72), we are solving a family of problems parametrized by the values of x_k. In most practical situations, iterating (eq.72) and (eq.73) is therefore only possible by tabulating $I_k^{opt}(x_k)$ over grids of points in the state space of *xk*. This can become prohibitive if the state x_k is too complex, for example if it is a real vector with many dimensions. This difficulty is called the curse of dimensionality.

The formulation above is fairly general, since we do not make any specific assumptions on variables, functions, and sets that appear in (eq.66)–(eq.68); they can be real numbers or integers, the sets can be defined by inequalities or by listing their elements, etc. The recursive optimization (eq.70)–(eq.73) covers all specific cases. However, a restrictive element in our formulation is the fixed, predefined number of steps K. This may be an obstacle if we want to solve, for example, problems of reaching certain points or sets in the plane or in 3D space, problems of traversing graphs, or problems with stopping conditions.

We shall now go through the derivation (eq.66)–(eq.73) aiming at a modification that would allow us to relax the above limitation. Since our recursive optimization follows backwards from terminal to the initial state, then in order to make the number of steps variable, we may fix the index label of the terminal state and vary the index of the initial state. Such a system of numbering, involving recursions between $I_k^{opt}(x_k)$ and $I_{k-1}^{opt}(x_{k-1})$, can be introduced easily. In view of this consideration, we often formulate dynamical optimization problems like in (eq.66)–(eq.68) and we understand number of steps K as a free rather than as a fixed parameter.

Nothing was said in (eq.66)–(eq.73), about admissible ranges of the states x_k, which may be of basic importance when one is programming practical solutions to dynamic programming problems. There may exist constraints on state variables in the problem formulation, in the form of inequalities or listing elements of sets, as it was mentioned above. In the context of optimization problems with the variable number of steps, the following recursive definition of admissible sets for states are often useful:

$$X_K = \{x_k : \text{iterations in (eq.66) terminate}\}, \qquad \ldots(74)$$

then

$$X_{K-1} = \{x_{K-1} : \exists_{u_{K-1} \in U(x_{K-1})} \text{ such that } f(x_{K-1}, u(x_{K-1})) \in X_K\} \qquad \ldots(75)$$

and, successively,

$$X_{k-1} = \{x_{k-1} : \exists_{u_{k-1} \in U(x_{k-1})} \text{ such that } f(x_{k-1}, u(x_{k-1})) \in X_k\}. \qquad \ldots(76)$$

Now, at the optimization stage k in recursive optimization (eq.70)-(eq.73) we include the condition $x_k \in X_k$. Again, we solve the parametric optimization problems (eq.72)–(eq.73), going backwards, i.e., for K, $K-1$, ..., with the additional constraint $x_k \in X_k$.

Tracing a Path in a Plane

Let us consider the problem, sketched graphically in Fig. 12.6, of tracing a minimal-cost path, in the plane, starting from a point P_0 and ending at a point P_K. The cost of a fragment of a path of length l is

$$C(l) = c_k l,$$

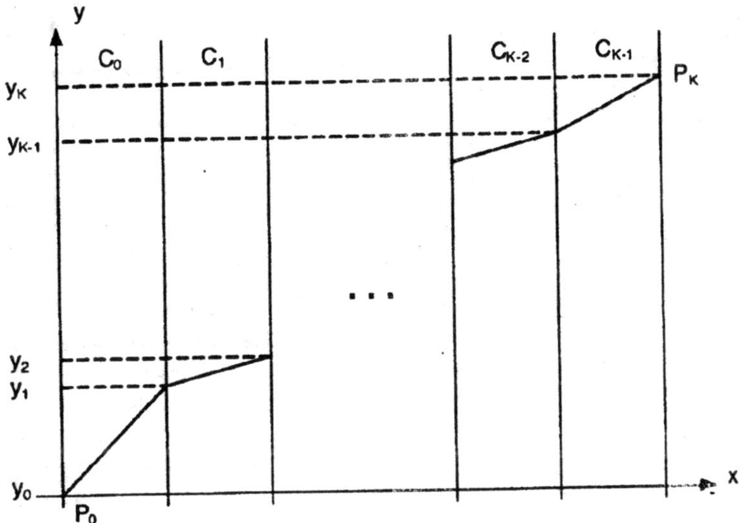

Fig. 12.6. Tracing a minimal-cost path from a point P_0 to a point P_K

where c_k is a coefficient, $k = 0, 1, \ldots, K-1$. The area between P_0 and P_k is divided into vertical strips, each of which has different cost coefficient c_k. The widths of successive vertical strips are denoted by d_k, $k = 0, 1, \ldots, K-1$. Clearly, if all the cost coefficients were equal, the optimal path from P_0 to P_K would be a straight line. Owing to the unequal costs in different strips, the optimal path is a sequence of straight-line segments and in order to arrange them optimally we can use dynamic programming. We denote the state of the discrete process of decision making at stage k by y_k, equal to the y-th coordinate of the optimal path when it crosses between strips $k-1$ and k. The recursion may be

$$y_k = y_{k-1} + u_{k-1}$$

where the value of u_k describes the change between two successive states. The cost of crossing the kth strip can be expressed as $c_k\sqrt{d_k^2 + u_k^2}$, and the scoring index for the optimization problem is then

$$I = \sum_{k=0}^{K-1} c_k \sqrt{d_k^2 + u_k^2}.$$

Since the path must hit the point P_k, we have the following constraint, which is one element set for $k = K-1$:

$$u_{K-1} \in U_{K-1}(y_{K-1}) = \{y_K - y_{K-1}\}. \qquad \ldots(77)$$

The controls u_0, u_1,..., u_{K-2} are not constrained. As we can see, the above is an instance of the formulation (eq.66)(eq.68) and the solution can be obtained recursively as in (eq.70)-(eq.73). Bellman's equation has the form

$$I_k^{opt}(y_k) = \min_{u_k}\left[c_k\sqrt{d_k^2+u_k^2}+I_{k+1}^{opt}(y_k+u_k)\right] \quad ...(78)$$

for k = 1,2,..., K^{-2}, and for k=K^{-1} we have

$$I_{K-1}^{opt}(y_{K-1}) = c_{K-1}\sqrt{d_{K-1}^2+(y_K-y_{K-1})^2}.$$

Even in this relatively simple case we are not able to compute an analytical solution. Instead, we approximate the possible range of y_k by a discrete set of for example, N = 1000 grid points, and we proceed by updating (eq.78) over the grid defined. Rigorously speaking, with this approach we obtain only an approximation to the solution to the formulated problem. We can improve the approximation merely by increasing N? If we want a solution with an accuracy as high as possible, an approach better than dynamic programming would be a variational formulation.

Shortest Paths in Arrays and Graphs

In Fig. 12.7, on the left-hand side, we present the problem of programming the optimal crossing through an array of numbers, with the aim of minimizing the score function given by the sum of the numbers in the cells of the array. The path starts at the bottom-left corner and ends in the top-right corner, and the feasible moves (decisions) are $\rightarrow, \uparrow$, and $\nearrow$. In this problem, the state of the process at stage k is

$$x_k = [x_k^r x_k^c] \quad ...(79)$$

where x_k^r and x_k^c are the indices of rows and columns of the array. We assume that the bottom-left corner of the array corresponds to numbers x^r = 1, x^c = 1 and the top-right corner to $x^r = R$, $x^c = C$. The state transition function is therefore

$$x_{k+1} = f(x_k, u_k) = \begin{cases} \left[x_k^r + 1 x_k^c\right] \text{for } u_k = \rightarrow \\ \left[x_k^r x_k^c + 1\right] \text{for } u_k = \uparrow \\ \left[x_k^r + 1 x_k^c + 1\right] \text{for } u_k = \nearrow \end{cases} \quad ...(80)$$

Denoting the scores in the cells in the array by

$$s(x^r, x^c),$$

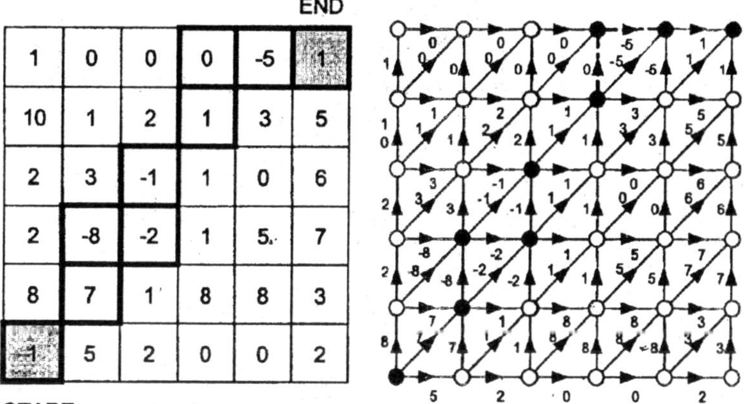

Fig. 12.7. Left: the problem of crossing an array of numbers from the bottom left corner to the top right corner, such that the sum of scores is minimized. Right: equivalent formulation as a graph-traversing problem.

for example, $s(2,1) = 7$, we can write the scoring index for the problem as

$$I = \sum_{k=1}^{K} s(x_k^r, x_k^c). \qquad \text{...(81)}$$

The number of steps K in the decision making process will depend on the path through the array. Bellman's equation takes the form

$$I_k^{opt}(x_k) \min_{u_k \in \{\rightarrow, \uparrow, \nearrow\}} s(x_k^r, x_k^c) + I_{k+1}^{opt}\left[f(x_k, u_k)\right] \qquad \text{...(82)}$$

and the optimal decision at stage k is

$$u_k^{opt}(x_k) - \arg \min_{u_k \in \{\rightarrow, \uparrow, \nearrow\}} s(x_k^r, x_k^c) + I_{k+1}^{opt}\left[(x_k, u_k)\right]. \qquad \text{...(83)}$$

The admissible sets for controls depend on states since the path cannot cross boundaries of the array, and so they reduce to $u_k \in \{\rightarrow\}$ if $x_k^r = R$ and to $u_k \in \{\uparrow\}$ if $x_k^c = C$. The optimization in (eq.82) is very simple, and proceeds by inspection of at most three elements. Also, since the states are discrete and finite, the values of $I_k^{opt}(x_k)$ are easy to tabulate. From (eq.82), we see that the optimal partial cumulative scores can be stored in a matrix of a size corresponding to the size of the array of scores $s(x^r, x^c)$. The order of filling in the entries of the matrix of optimal partial cumulative scores must be such that (eq.82) is always manageable. Also, in the course of computing the optimal partial scores, one can record the optimal decisions. We can trace the optimal strategy by following arrows in

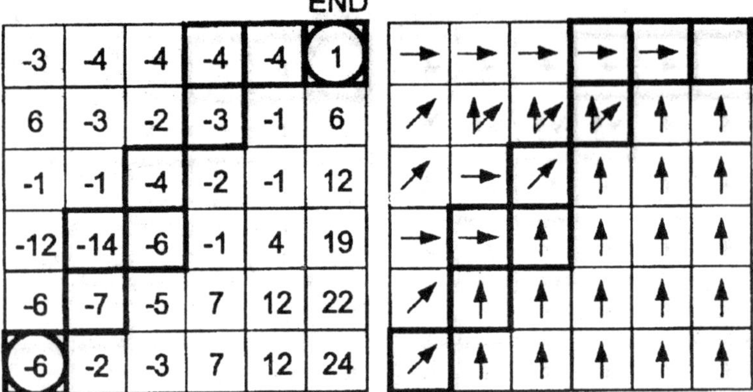

Fig. 12.8. Left: matrix of optimal partial cumulative scores for the problem in Fig. 12.7. Right: matrix of optimal controls. The optimal path is depicted by bold squares.

this plot, and we see that there are two diûerent paths of equal score $I_1^{opt}(x_1) = -6$. It is also possible to find an optimal path directly from array of optimal partial scores, without recording optimal decisions, by following the "*steepest descent*", or, in other words, by repeating (eq.82) for the array of optimal partial scores. Both the array of optimal partial cumulative scores and the array of optimal decisions allow us to find optimal path to the final state not only from the initial cell $x^r = 1$, $x^c = 1$ but also from any other cell in the array.

Graphs and shortest paths

The problem described above is very closely related to planning shortest paths in graphs. In the right part of Fig. 12.7, a graph is presented, that is equivalent to the array of scores in the left part. Each node corresponds to one cell of the array and each arrow corresponds to a feasible transition between cells. All arrows are directed, and their lengths (weights) correspond to the scores in the cells of the array. The problem of scheduling an optimal path through an array, discussed above, is equivalent to that of designing an optimal path in a graph, with the minimal sum of weights. Clearly, it can be solved by a dynamic programming method analogous to that already described. What makes the solution relatively easy, and can be adequately formulated in terms of graph terminology, is aperiodicity. A directed graph is aperiodic if, after departing from any of nodes, there is no possibility of returning. Clearly the graph in the right part of Fig. 12.7 has this property. For an aperiodic graph there is always a method to assign integer numbers to the nodes such that if, for nodes x and y, number(x) < number(y), then there is no path from y to x. The node with the smallest number will not have any entering vertices (arrows), and the node with the largest number will not have any exiting vertices. Using numbering of nodes, one can easily order the optimizing decisions in an appropriate way and therefore efficiently solve for the shortest path. What if the graph is not aperiodic? In this case it can have cycles; after departing from some node, there might be a possibility to return after traversing some other vertices. For example, if we assume that the possible moves in the array of scores in Fig. 12.7 are now $\rightarrow, \uparrow, \nearrow, \leftarrow, \downarrow, \swarrow$ then the corresponding graph can obviously have cycles. Planning paths for graphs with cycles is more difficult than in the aperiodic case. It is also necessary to introduce some conditions on scores (weights), for example, $s(x_k^r, x_k^c) \geq 0$, when minimizing the total score of the path. Without this condition it could happen that a cycle has a score that is negative and one can make the total score go to $-\infty$ by performing loops around this cycle. Actually, we could observe this in the formulation in Fig. 12.7 if we allowed moves $\rightarrow, \uparrow, \nearrow, \leftarrow, \downarrow, \swarrow$. An efficient algorithm for solving for the shortest path in general graphs was formulated by Dijkstra; this algorithms can be stated with the use of dynamic programming.

Combinatorial Optimization

Combinatorial optimization problems are commonly understood as optimization problems over mathematical objects such as paths, trees, graphs, such that their listing, labeling or enumerating involves using combinatorics. Combinatorial optimization problems can be hard to solve owing to the large solution space, which is difficult to explore. More precisely, combinatorial optimization is a branch of optimization theory dealing with the complexity of optimization and decision problems and the related classification of algorithms for solving optimization problems. Combinatorial optimization has links to branches of computer science and applied mathematics, such as algorithm theory, artificial intelligence, operations research, discrete mathematics, and software engineering. Knowledge and experience in the field of the computational complexity of an instance of an algorithm becomes critically important when the size of the problem increases. Since in bioinformatics the size of the data, i.e., sequences and measurements, is usually very high, exploring the computational complexity of the algorithms is very important. In this section, we overview the classification of optimization or decision problems from the point of computational complexity and give examples of combinatorial optimization problems. Excellent presentations of the present state of the art in combinatorial optimization can be found in the monographs.

Examples of Combinatorial Optimization Problems

We start by presenting several examples of combinatorial problems.

Traveling salesman problem

For every pair out of K cities $C_1, \ldots, C_K$, we know the distance or the cost of travel between them, $d(C_i, C_j)$. The problem is to find the shortest (or cheapest) route through the cities $C_1, \ldots, C_K$, such that each of the cities is visited at least once.

Hamiltonian path problem

Given a graph G, verify whether there exists a Hamiltonian cycle for G. A Hamiltonian cycle is a path along the edges of a graph such that every vertex (or node) is visited exactly once.

Shortest-superstring problem

Given a collection of words $w_1, \ldots, w_K$ over an alphabet, find the shortest string that contains all words $w_1, \ldots, w_K$.

Boolean satisfi ability problem

Given a Boolean function (expression) $f(b_1, \ldots, b_K)$ over binary variables $b_1, \ldots, b_K$, determine whether we can assign values, zero or one, to each of the variables $b_1, \ldots, b_K$ such that the Boolean formula $f(.)$ is true, in other words, that will set $f(b_1, \ldots, b_K) = 1$.

Time Complexity

We can assign a size to a combinatorial optimization problem. In the problems listed above, the size is given by the number K. The size is proportional to the length of the data string fed to the algorithm for solving the problem.

By the time complexity of a problem or of an algorithm for solving an instance of a problem, we mean the relation between the running time of the algorithm and the size of the problem. More formally, the "running time" can be replaced by the number of steps required by a Turing machine programmed for the execution of the algorithm.

Decision and Optimization Problems

There are two types of problems, decision problems (determine whether an object with given properties exists or not) and optimization problems (find the object which optimizes a criterion, i.e., the cheapest, shortest, etc.). However, we can demonstrate that the distinction between decision and optimization problems is not very important from the point of view of their time complexities. For example, let us replace the traveling-salesman optimization problem stated above by the following traveling salesman decision problem: Decide whether there is a route visiting each of the cities $C_1, \ldots, C_K$ at least once and such that its total cost is $\leq \theta$, where θ is a given number. Assuming that we have an algorithm for solving the traveling salesman decision problem, we can repeat this algorithm several times and use the idea of bisection of an interval to obtain reasonable knowledge about the optimal route. Roughly, the number of repetitions of the decision algorithm necessary will be proportional to $\log_2$ (size). So, having an algorithm of time complexity Time(size) for solving the traveling salesman decision problem we can design an algorithm for solving traveling salesman optimization problem with time complexity $\log_2$ (size) $\times$ Time(size). If a combinatorial problem belongs to one of the classes polynomial or exponential, then multiplying it by $\log_2$(size) does not change the class. Therefore, for the traveling salesman problems, optimization and decision problems belong to the same class. An analogous argument can be applied to other combinatorial problems.

Classes of Problems and Algorithms

The classification of problems is related to their time complexities. The class P includes problems for which there are algorithms with a polynomial time complexity. The classes NP, NP-complete, and NP-hard include problems whose time complexities are most probably higher.

Let us present the classes NP and NP-complete more precisely. On the basis on equivalence between optimization and decision problems demonstrated above, we focus only on decision problems, which have the property that the output of the related algorithm is yes or no. The name "NP" is an abbreviation for "*nondeterministic polynomial*". Problems that belong to this class have a polynomial-time certificate. A certificate here is an algorithm used to determine whether a decision guess satisfies a condition. For example, in the traveling salesman problem, we may construct (guess) any route through all the cities $C_1, \ldots, C_K$. The existence of a certificate means that, in polynomial time we can find whether the proposed route satisfies "cost $\leq \theta$" or not. NP problems can be solved by a nondeterministic Turing machine. A nondeterministic Turing machine is a Turing machine additionally equipped with a guessing, write-only head. The class NP clearly includes all P-problems, $P \subset NP$, since they can be not only certificated but also even solved in polynomial time. Among the problems in the class

NP there is a subclass, called NP-complete. Problems in the class NP-complete have the property that any problem in the class NP can be reduced to a problem in the class NP-complete in polynomial time. The first result concerning NP-completeness was Cook's theorem, stating that the Boolean satisfiability problem was NP-complete. Following Cook's theorem, many other problems have been proven to belong to the class NP-complete.

One more class of combinatorial problems is the class named NP-hard. The class NP-hard contains all problems H such that every decision problem in the class NP can be reduced to H in polynomial time. The difference between classes NP-hard and NP-complete is that for NP-hard problems we do not demand that they must belong to NP. In other words, these problems may not have certificates.

Suboptimal Algorithms

An important field is the development of combinatorial algorithms for NP- complete problems, called suboptimal, near-optimal, or approximate. These algorithms are of significantly lower complexity; most often they work in polynomial time. Despite the fact that they do not guarantee that the solution will be obtained but only that one will get close to it, the results they provide can be acceptable and useful in many practical applications. Some examples, also mentioned later, are polynomial algorithms for suboptimal solutions of the shortest-superstring problem and polynomial algorithms for approximate solutions of the Hamiltonian path problem.

Unsolved Problems

Combinatorial optimization and decision problems have been studied extensively. The research in this area has two main directions. The first involves improvements in performances of algorithms. If the best known algorithm for solving a specific problem has a time complexity $C \exp(K)$, where K is the size of the problem, then developing an improvement leading to time complexity $(C/2) \exp(K)$ may be a useful and publishable result. Developing an algorithm, which improves the time complexity from $O(K^2)$ to $O(K \ln K)$ is a substantial advance, which can result in the appearance of new methods and new applications in related areas.

The second involves proving results concerning classification of problems. As already stated the classification for large number of problems, according to the above rules, has been established. Establishing time complexity classification of many problems, for example of the linear programming problem or the problem of factorization of an integer, involved many years of research.

Combinatorial optimization and the theory of algorithms contain many unsolved problems. First, the class of many NP problems is unknown; they have not been neither proven to belong to the NP-complete class, nor has a polynomial time algorithm been found. Moreover, the famous hypothesis P = NP is still unsolved. The common belief is that classes P and NP-complete are disjoint. But nobody has proven that any problem from the NP-complete class cannot be solved in polynomial time.

INDEX